UPSHIFT

ALSO BY BEN RAMALINGAM

AID ON THE EDGE OF CHAOS

UPSHIFT

TURNING **PRESSURE** INTO **PERFORMANCE** AND **CRISIS** INTO **CREATIVITY**

BEN RAMALINGAM

FLATIRON
BOOKS
NEW YORK

www.flatironbooks.com

The Library of Congress Cataloging-in-Publication Data is available upon request.

ISBN 978-1-250-79295-2 (hardcover)
ISBN 978-1-250-83194-1 (ebook)

Our books may be purchased in bulk for promotional, educational,
or business use. Please contact your local bookseller or the Macmillan
Corporate and Premium Sales Department at 1-800-221-7945, extension 5442,
or by email at MacmillanSpecialMarkets@macmillan.com.

First Edition: 2023

10   9   8   7   6   5   4   3   2   1

For Koby, who still thinks this book
would be better if it were called
Innovators in a Pickle

CONTENTS

PART I: THE SCIENCE OF UPSHIFT

INTRODUCTION   3

1. THE UPSHIFTING MENTALITY
Getting the Butterflies to Fly in Formation   21

2. ORIGINALITY UNDER PRESSURE
Great Minds Think Differently   45

3. THE STRENGTH OF PURPOSE
What Can Never Be Taken Away   81

4. THE POWER OF UPSHIFT
A Reprise   101

PART II: THE ART OF UPSHIFT

INTRODUCTION   117

5. CHALLENGERS
Change the Rules   121

CONTENTS

6. CRAFTERS
Struggle for Novelty   149

7. COMBINERS
Remix Across Boundaries   175

8. CONNECTORS
Harness Network Intelligence   199

9. CORROBORATORS
Assemble the Knowledge   227

10. CONDUCTORS
Orchestrate Change   253

EPILOGUE
The Master Switch   281

ACKNOWLEDGMENTS   285

NOTES   287

DOWNSHIFT

1. A change to a lower gear in a vehicle or bicycle.

2. The triggering of the distress mode function in the brain when something is perceived as a threat, real or imagined. Associated with a neurological shift from higher cognitive processing functions into more basal regions associated with fight, flight, or freeze responses. Often associated with raised heart rates, heightened adrenaline levels, and a sense of anxiety.

UPSHIFT

1. A movement of a variable to a higher level, e.g., of performance, growth, frequency.

2. The triggering of the eustress mode function in the brain when something is perceived as a challenge, real or imagined. Associated with a shift into higher cognitive processing regions that facilitate novel ideas, associations, relationships, and solutions.

PART I

THE SCIENCE OF UPSHIFT

INTRODUCTION

CLICK

"Come on, *thamby*," Mum says. "It won't be long, and we are going where we will be safe from the fighting."

Safe. It is the word that everyone uses, all the time, to explain everything. When our dog Veera wanted to come with us, I had to tell him no. But he wouldn't listen because he *knew* we were going. The last time I saw him, he was cowering, tail down and whimpering. Mum had to close the door because I couldn't.

"We won't be *safe* if he comes, *thamby*," she had said, wiping away my tears.

When I hugged my grandmother, she whispered in my ear, "This is my home and my country, and not all the stupid fighting sons and nephews and grandsons in the world will take me away from it. Remember that."

"*I'm* not fighting," I said, pulling away, stung by her inclusion of "grandsons." "And I never will."

She laughed then and pinched my cheek, but not hard. "You will not have to, because you are going where it is *safe*."

Safe. We had to leave our home at night, so we would be *safe*. We

3

had to go across the smelly mudflats to another house that was more *safe*. We had to leave town—maybe even the country—to go where it was *safe*. Daddy would go on ahead to make sure it was *safe*. We had to leave Veera behind, so we would be *safe*.

There were all these rules for being *safe*. And only the grown-ups knew them. And they kept changing them all the time. Being safe was slow and boring and meant lots of waiting. Until being safe suddenly happened very fast and was painful and scary. And then being safe was back to slow and boring.

Behind us, a young man jumps up and shouts, *"Katamarang!"*

I and the dozens of others waiting turn to the lagoon to see the barge approaching in the distance. Everyone is getting up excitedly. The barge chugs toward us, puffing smoke. As it gets closer to the dock, the mixture of steam and petrol in the air smells beautiful and free to me. The dockman starts pulling out the boarding planks.

"Stay all where you are," a voice barks through a loudspeaker in English. Four soldiers in a jeep screech to a halt by the dock. Two of the soldiers get out of the jeep with rifles. They start moving through the crowd, searching people.

The soldiers pass by us and look in our bags. I can smell cigarettes on them, foul and nasty. I want to shout out, or punch out, or jump up, or run. Things are happening fast again. I want to outpace them. My mum puts an arm over me and breathes quiet words as the soldiers move past us.

"Be still, *thamby*. Look down. Remember where we are. Don't give them any reason."

I heard about where we were on one of my last days at school. Many people had been shot dead on the lagoon. Boats full of dead bodies—men, women, and children—and soldiers pushing them into the water with their feet. As I look down at the surface of the water, I imagine their faces looking up at me.

The barge has a slow, gentle rise and fall, up and down with the swell of the lagoon, like it is a breathing animal. I copy it with my own breathing, like I used to with Veera, and it makes me calm.

As we file on, the soldiers stand, rifles in hand, watching us board. I don't look back at the people who are not allowed on as we cross the boarding planks. I don't think about them.

Then, once the last ones are on, the soldiers also get on the barge. The passengers—my family too—move like sheep being herded: all at once, to get as far away as possible from the soldiers and their weapons. When the barge starts moving, the two soldiers stand opposite us, taking up an entire half, while we are all jammed into the other half. Like Veera, cowering.

The soldiers start to smoke cigarettes. Even their lighters look scary. I had heard those stories too; about what soldiers did with lighters. But as they smoke, I'm watching their cigarettes wobble, and I realize their hands are shaking. *They* are scared too. Despite the guns, they are scared.

Something in me goes *click*. I think: "Mum is scared. The people with us are all scared. The soldiers are scared. But I don't have to be scared. And if I'm not scared, neither should anyone else be. We can *all* feel *safe*."

I shake my mother's arms off and start to walk over to the soldiers from our hemmed-in group. My mum lets out a strange, terrified whisper-scream. Others in the group call me back. But I don't look back. I carry on, putting one foot in front of the other. The soldiers suddenly look up from their talking and smoking, startled to see me so close.

"Can I see your gun?"

They stare at me. I ask again, pointing. And they start to laugh.

"I have a nephew your age," one says.

"I'm eight," I say.

"My nephew is nine now," he replies. "I haven't seen him in . . ." He trails off and smiles at me.

"Am I taller than your nephew?" I ask.

"Yes," he says, "you are very tall."

The younger soldier is about to light another cigarette, and the one talking to me says, "Not in front of the boy."

They look at each other and both start to laugh again. One of them reaches out and ruffles my hair. We carry on talking. Behind me I hear more talk and laughter. I look back and all I can see is my mum smiling and crying at the same time.

Everyone relaxes when they see me having fun with the soldiers. Passengers move around, talking to one another. The barge becomes more sociable, relaxed. When we all disembark, the soldiers wave goodbye to me and my family, and we carry on with our journey to safety.

I'm well aware that this experience could have gone very differently. Many accounts of the brutality of the civil war in Sri Lanka—at that time, in that location, on those vessels—attest to how fortunate we were. And I'm not for a moment suggesting that I was somehow responsible for changing the course of events. My experiences in conflict zones since then have taught me that a single moment of empathy is seldom enough to change the orders or intentions of warring parties bent on violence.

But my childhood experience does pose several questions, and as I will show in the pages ahead, they lead to one I have explored in many different ways throughout my life: Why did I break free from my mother's "safe" arms? Why did I ignore the pleas to come back? What compelled me to walk up to an armed soldier and ask to see his gun?

What actually happened in that moment, in that *click*?

THE INVERTED U

I have learned that click moments such as the one I experienced can be found in all situations—from pilots landing downed aircraft to commuters navigating disruptive incidents to children play-fighting. Every single one of us will be able to identify click moments of our own.

Take work, just by way of illustration. If I asked you to think about times when you have experienced overt pressure in your job, you are likely to groan and think of some or all of the following: overbearing bosses, demanding clients, ineffective or insensitive colleagues, apathetic junior staff, grumbling partners, and impossibly slow suppliers.

Imagine I then asked, when has pressure been *useful* at work? I bet we will all remember some bright spots. Perhaps having to work on a thorny problem, late at night, and suddenly reaching a breakthrough idea that made sense of the whole project. Maybe you had a group of colleagues whose differences in perspective initially triggered frustration and anger, but then you found an unexpected form of words that made everyone shift into mutual respect, and a memorable working partnership resulted. Once you start, you will be able to identify many examples in your working life where pressure had an upside and the click moments that got you there.

In fact, there is a wealth of research on the subject of when stress and pressure might be good for performance. Examples cover the full spectrum of human experience. From gymnasts in China to interior designers in Germany; from airport staff in Ghana to soldiers in the US; from market traders in Uganda to clinicians in the UK. These real-world observational studies, spearheaded by psychologists, anthropologists, and management scholars, have been reinforced by many hundreds of lab-based experimental assessments. "When is stress useful?" has been one of the most significant questions in individual and group psychology in recent years. It has been asked so many times, in so many settings, that in seeking a rigorous and credible answer I have been able to collect evidence covering the work of thousands of individuals and teams from around the world.

Now think back to those unproductive work stresses. Most likely you will remember feeling under threat, resorting to risk-averse and rigid thinking, being demotivated and losing your sense of purpose. The more stress you face, the worse these things get. At their very worst, as many of us have learned first-hand, these experiences can diminish our healthy functioning, and not only at work. Even before COVID-19, workplace stress was seen as being at near-epidemic levels. In the UK alone, over fifteen million working days were lost due to stress in 2018.

But this is only part of the story. It is equally well understood that with *too little* stress and pressure, we become disengaged, de-motivated, and unfulfilled. The Midlife in the United States investigation, a nationally representative study that has been running since 1995, measures psychological and social factors that influence health. One of the most fascinating workplace-related findings was that the lack of stimulation in work has a long-term cognitive effect on employees. This may seem counterintuitive: we assume that as we grow older we should challenge ourselves less. Many of us naturally seek out less complex and stressful tasks and occupations as a means of protecting our state of mind and general well-being. But in fact the opposite happens: the lack of stimulation and productive stress actually diminishes our mental and biological well-being and lowers our peak performance.

Psychologists have been investigating the effects of both too much and too little stress for well over a century. In 1908, Harvard psychologists Robert Yerkes and John Dodson designed an experiment that tested the positive and negative effects of stress on performance. Today their work is referred to as the Yerkes-Dodson law, represented by the inverted U curve shown on the following page.

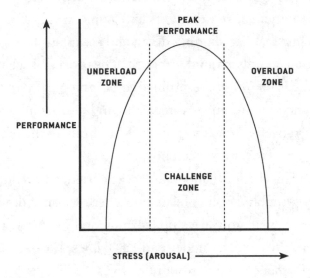

The Midlife study found this relationship to hold true across all participants: if we put ourselves in positions where we continually have to learn new skills and take on new challenges, this results in stronger cognitive performance—and this becomes especially marked as we get older. The same results have been reinforced by thousands of neuroscientific studies of so-called "super agers," who perform mentally and physically at levels comparable to people half their age: the secret trick is to regularly push yourself beyond your comfort zones both physically and mentally.

This is very clear evidence that pressure and stress can be vital for our well-being. The same phenomenon has been identified in many professions—fighter pilots, engineers, medical staff, police officers—and in many day-to-day activities—playing sports, speaking in public, even participating in weight-management programs.

Thanks to its ubiquity, the inverted U has become popularized, developed, built upon, and used extensively in neuroscience, psychology, and medicine.

The right-hand side of the curve represents the negative effects of stress. When we experience too much stress, we are overloaded. We feel it's hard to organize our thoughts and gain control of the situation. We can just feel like shutting down and escaping the source of stress. This is an entirely natural response. The process is what brain scientists call *downshifting*: the protective neurological dampening that occurs when you're under threat, moving your thinking from the cognitive, reflective, creative parts of your brain to the more primeval areas geared toward survival.

By contrast, on the left-hand side, we experience boredom, apathy, and lack of motivation. This is not where we shut down, but rather where we switch off; we stop learning and growing. In fact, if we experience too much "underload" as babies, it can affect our cognitive performance for the rest of our lives.

In between the two is the sweet spot where we experience what psychologists have called *eustress*.

We start to move toward this sweet spot when we switch from perceiving a stressful situation as a challenge rather than a threat. This is where we start to move from the left- or right-hand below-par-performance sides into the peak-performance zone. We *click* into being at the top of our game: into Upshifting.

MENTALITY, ORIGINALITY, PURPOSE

When we experience a click moment, our brains move into a higher cognitive process, and we are able to come up with novel ideas, associations, relationships, and solutions.

You might say, "Aha, necessity is the mother of invention!" While this is indeed part of the story of Upshifting, it is an imperfect and often misleading aphorism. Evidence and experience suggests that most of the time, the pressure of necessity actually leads us to *convention*, not invention. When we are under pressure, most of us gravitate toward the safe, the tried and tested. A more accurate (and less aphoristic) statement would be "necessity—under certain conditions, and with essential ingredients—can lead to invention."

Upshift is an exploration of how we can turn pressure into performance and crisis into creativity. In this book, we're going to go around the world and find out about people in every walk of life and how they have reached and capitalized on the Upshift sweet spot.

We will learn that there is an underlying pattern to how all of these people have converted necessity to invention—how they have dealt with and harnessed pressure and stress and how they have responded to crises through creativity.

Necessity does indeed provide the stimulus, the catalyst for the click. But I have found that capitalizing on necessity requires three essential ingredients. When we have them all, we Upshift.

These same three things apply to parents managing weekly budgets, office workers navigating workplace stresses, and disaster responders working in war zones. I have even used them to better understand my actions as an eight-year-old boy on that barge.

If you think back to your own experiences of productive stress at work, you probably remember having the sudden feeling that the pressure could perhaps be capitalized upon, that it was a challenge and not a threat.

In my memory, it was the shaking of the cigarettes held by the soldiers that made me realize that although they were different from us (soldiers, armed, in charge), they were also like us because they were scared.

It was the wobbly cigarettes that made me click from the *mentality* of facing a distressing threat to facing a stimulating challenge. As we will see, when a threat is reappraised as a challenge, we reevaluate the whole situation and its limitations, and the possibilities that might unfold.

In this way, click moments open the cognitive and emotional door to Upshifting. But we still have to walk through. This means harnessing the *originality* of ideas and approaches that emerge under pressure and stress.

That soldier didn't get on the barge thinking he would engage with one of the displaced civilians—and a child, at that. On one level, it was an innocent interaction that belonged to a prewar era. At a certain point during our conversation, we all started to see our situation from a different perspective, the possibility of being people being together, being safe. The situation had changed into something utterly unexpected.

Finally, the Upshift of pressure into performance, of crisis into creativity, depends on a shared *sense of purpose*.

The soldiers and I could not have been more different. They had orders, were on the front line of a civil war, and were trained and experienced. I was young, a member of the opposing ethnic group, innocent. The soldier's own nephew bonded us. A similar age, a similar height, and—although I have no way of knowing this—perhaps a similar cheekiness.

Looking back, the moment when he stopped his companion from smoking in front of me was the first step to a sense of shared purpose: the recognition that my safety mattered. How we then talked and laughed together, a process of developing a feeling of mutual safety between us, which then spread to everyone else on the barge.

These are the three ingredients of what I now understand as Upshift: *mentality*, *originality*, and *purpose*.

As a child, I believe I Upshifted several times, and those occasions are among my most intense memories of growing up. Most are ordinary examples: teaching myself to ride a bicycle because of absent adults, or overcoming my older brothers' unkind school bullies. But the episode on the barge stands out. Among my childhood memories, it is one of the few where I can remember exactly what I thought and how it felt: an extended experience not viewed through a vague mist, but captured and crisply delineated.

My childhood experience of living through and fleeing from the Sri Lankan conflict has loomed over and shaped some of my most important life decisions. In my early twenties, on a journey across India with one of my best friends, I met Sri Lankan refugees who had left at the same time my family did almost two decades previously, but were still living in a tin-hut slum on the edge of Chennai. It made me realize how lucky we were, despite all the troubles we had experienced as a family fleeing from war in a foreign country.

I spent days in a V-shaped depression before coming to the realization that I had to try to help people who were suffering from the same kinds of situations my family had managed to escape—albeit not unscathed. Within three years, my professional life became reoriented to work on conflicts and disasters.

I have now spent two decades working with and advising organizations like the Red Cross, the United Nations, and Doctors Without Borders (Médicins Sans Frontièrs). I have worked to bring innovation and creativity into the international responses to extreme events.

This has led me to many corners of the world where tragedy has struck: tsunamis in the Indian Ocean and Japan; earthquakes in Pakistan, Haiti, and Nepal; cyclones in Bangladesh and Myanmar; floods in Indonesia; conflicts in Gaza, Sudan, the Democratic Republic of the Congo, Afghanistan, Syria, and Ukraine; epidemics in Haiti and West Africa; and the ongoing coronavirus pandemic.

What I have learned—like many others working in crisis response—is that extreme events stretch the boundaries of what is possible. They create huge losses and unspeakable heartbreaks, it goes without saying. But amid devastation and destruction, time and again, I saw many different kinds of people demonstrate the mentality, the originality, and the purpose that meant they could do more, save more, and restore more. In writing this book, I moved beyond crisis response to examine other contexts—from the military to space travel, from sports to arts. And I have seen the same ingredients—and click moments—in every area of human endeavor.

Which brings me to one of the most renowned examples of Upshifting in recent years.

CAPTAIN SULLY'S CLICK MOMENT

At 1524 Eastern Time on January 15, 2009, the tower controller at New York's LaGuardia Airport cleared US Airways Flight 1549 for takeoff from runway 4. The plane, an Airbus A320 captained by Chesley "Sully" Sullenberger, took off to the northeast. After about two minutes, at around 3,200 feet, First Officer Jeff Skiles noticed a large formation of birds in the skies to his right.

At 1527, the radar showed the flight being struck by what turned out to be a flock of wild geese. In the cockpit, the windscreen turned dark brown and loud crashes reverberated through the aircraft.

What follows is a simplified extract from the transcript of the cockpit recorder of the sixty seconds that followed the impact, including my own interpretation of when Sully had his click moment. The plane's call sign is Cactus 1549 (the variations on this—1529, 1539—were errors made by those speaking).

> **TRAFFIC CONTROLLER #1:** Cactus 1549, turn left plane 270.
>
> **CHESLEY SULLENBERGER:** Ah, this is, uh, this is Cactus 1539. Hit birds. We've lost thrust in both engines. We're turning back towards LaGuardia.
>
> **TRAFFIC CONTROLLER #1:** Okay, yeah, you need to return to LaGuardia. Turn left heading of, uh, 220.

SULLENBERGER: 220.

TRAFFIC CONTROLLER #1: Tower, stop your departure. We got an emergency returning.

TRAFFIC CONTROLLER #2: Who is it?

TRAFFIC CONTROLLER #1: 1529 he—ah—bird strike. He lost all engines. He lost the thrust in the engines. He is returning immediately.

TRAFFIC CONTROLLER #2: Cactus 1529—which engines?

TRAFFIC CONTROLLER #1: He lost thrust in both engines, he said.

TRAFFIC CONTROLLER #2: Got it.

TRAFFIC CONTROLLER #1: Cactus 1529, if we can get it to you, do you want to try to land runway one three?

SULLENBERGER: We're unable.

TRAFFIC CONTROLLER #1: All right, Cactus 1549, it's going to be left traffic to runway three one.

SULLENBERGER: Unable.

TRAFFIC CONTROLLER #1: Do you want to try and go to Teterboro?

SULLENBERGER: Yes.

TRAFFIC CONTROLLER #1: Cactus 1529, turn right 280. You can land runway one at Teterboro.

SULLENBERGER: We can't do it.

TRAFFIC CONTROLLER #1: Okay, which runway would you like at Teterboro?

Click!

SULLENBERGER: We're gonna be in the Hudson.

TRAFFIC CONTROLLER #1: I'm sorry, say again, Cactus. . . . Cactus 1549 Radar contact is lost . . .

TRAFFIC CONTROLLER #3: I don't know. I think he said he was
going in the Hudson.

After the point when the radar contact was lost, there was no further communication between Sullenberger and the traffic controllers. At 1530, Flight 1549 splashed down in the Hudson River in
what was subsequently referred to as the most successful ditching
in aviation history. After all the cross-checks, and a rapid evacuation of passengers onto the wings of the plane, Captain Sullenberger
walked up and down the length of the plane twice to check that no
one was left on board.

There is a huge amount that has been written about, and by, Sullenberger, and about those three and a half minutes. In one of the countless interviews he gave after the rescue, he described three things he
did that, in his words, "made a great deal of difference" to the outcome.

> My body's initial physiological response to the startle factor was
> huge. I was aware of it in the first seconds as it happened. My
> blood pressure and pulse spiked. I got tunnel vision; my percep-
> tual view narrowed because of the stress. It was marginally de-
> bilitating, it really interfered with my mental processes.

Captain Sullenberger's immediate action was to calm himself and
get into the mentality of being able to meet the challenge, and this
despite his body screaming "threat!"

> First, [I had to] force-calm myself. The kind of professional calm
> that we learn to summon up from somewhere within us, which
> isn't so much the calm as having the discipline to compartmental-
> ize our thinking and focus clearly on the task at hand . . . And that
> was difficult to do, it required a lot of effort.

In another interview, he spoke of his confidence at this point:

> I knew I could find a way. Even though this was an unanticipated event for which we have never specifically trained, I was confident that I could quickly synthesize a lifetime of training and experience and adapt it.

The next thing he did was seek out an *original* approach to the problem:

> We were suddenly confronted with the challenge of a lifetime . . . one we had never seen before. . . . We had only 208 seconds, just under three and half minutes, from the time we hit the birds and lost thrust in engines until we had landed, to solve this problem we had never seen before . . .
>
> Even though this is not something we are specifically trained for, based on my training and experience, I [had] to quickly take what I did know, and apply it in a new way to solve this problem.

Then he had to work with absolute focused *purpose*:

> The third was to make my workload manageable. I chose to do only the highest priority items, but do them very, very well. And then I had the discipline to ignore everything else that I didn't have time to do that would only be a potential distraction.

As he expanded in another interview:

> We had to have the presence of mind . . . to never give up, to always use all the resources to keep on trying to solve it, make it better, somehow. There's always one other action you might be

able to take that might increase our success by a little bit. And so, when, when you look at the cockpit voice recorder transcript and you see it, I said to Jeff, right before landing, "Got any ideas?" Some think that that was a flippant remark, but of course it's not. And Jeff understood in that context exactly what I was asking him, which was, 'I've done everything I can think of to do that would help us. Is there anything else you can think of that we can do that would help us be more successful by a fraction?' And his answer was actually not. And he answered that way, just in a very casual matter-of-fact way, not because he was resigned to an electable fate—far from it. We were trying to save every life to the very end. He answered that way because at that point, he knew we had done all we could.

Sullenberger described these three things as the things he did that made the *most difference* to the outcome. His account independently and powerfully verifies the three essential ingredients of Upshifting.

UPSHIFT: THE "WHAT" AND THE "WHY"

Drawing on my direct experience and extensive research, I have written this book to show how we can all recognize Upshifting, understand it, and, above all, learn to harness it.

I've watched people do it, I've worked alongside people and been buoyed by their Upshifts, I've been part of teams and groups that have Upshifted, and I have even supported and coached people to do it.

For many of the people we will learn about in the pages to come, the Upshift I describe was a peak moment in their lives. The resulting successes in their personal and professional lives shaped how they thought about themselves, their lives, and their work, often with positive ripple effects for those around them. Some have even sought to turn the process into a repeatable practice.

This book is my attempt to acknowledge what I have seen and heard, to share the stories of the incredible people I have been privileged enough to learn from, in some of the toughest situations in the world.

I have sought to tell the stories of how they have Upshifted not just as homage to their achievements, but also in the hope and belief that understanding their experiences will help you follow your own path toward Upshifting. This book contains all the lessons I have learned about how to repeat the trick of Upshifting—to move it from something akin to a dice roll to something we can all learn and improve, a skill and a capability.

This is the *what* of this book. As to the *why*, I have become increasingly convinced that these ideas and principles are of relevance to people in all walks of life, facing all kinds of challenges. If we are to not just survive but thrive in the era of perpetual crisis in which we find ourselves, we need to individually and collectively Upshift to new ways of working and living.

As Abraham Lincoln famously put it: "The dogmas of the quiet past are inadequate to the stormy present. The occasion is piled high with difficulty and we must rise with the occasion. As our case is new, we must think anew and act anew. We must disenthrall ourselves." This is as close a statement as anything to the core mantra of *Upshift*.

In his classic book on the psychology of happiness, *Flow*, creativity scholar Mihaly Csikszentmihalyi observes, with typically razor-sharp precision: "Of all the virtues we can learn no trait is more useful, more essential for survival, and more likely to improve the quality of life than the ability to transform adversity into an enjoyable challenge. To admire this quality means that we pay attention to those who embody it, and we thereby have a chance to emulate them if the need arises."

In exploring and documenting the relationship between pressure and performance, between crisis and creativity, I hope to have developed a field guide to exactly those essential virtues, traits, and abilities that Csikszentmihalyi identified, by paying careful and considered attention to those remarkable people who embody them.

The guide is in two parts. The first describes the three ingredients of Upshifting in more detail: mentality, originality, and purpose. The second describes six different styles of Upshifting I have discovered, and will help you understand your own styles as well as those of others around you.

My fundamental aim in *Upshift* is to challenge the assumption that stress and pressure kill performance and creativity, showing instead how there are in fact liberating opportunities to be found even in the most extreme of conditions. In doing so, I want to make the case for two vital changes.

First, I want to help to democratize the idea of performance under pressure as a capability that we all possess, and that we can all learn to improve.

Second, I want to help expand our understanding of the situations and contexts in which we can apply these capabilities.

To get there, I want to start with a broken piano.

1

THE UPSHIFTING MENTALITY

GETTING THE BUTTERFLIES
TO FLY IN FORMATION

THE SOUND OF ONE MAN JAMMING

On Friday, January 24, 1975, Keith Jarrett sat down at the piano on the main stage at the Cologne Opera House, in front of a packed house, and started to play. The concertgoers remember the moment well. In the words of one, speaking more than thirty years later, "I just remember seeing a single man on stage with a single instrument. It was just not normal at the time." The sense of wonder at the contrast between this solitary figure and the typical opera house concert, brimming with players and singers, is still clear in the voice of the gentleman in question.

And yet this concert, which would go on to make musical history, almost didn't happen at all. The buildup had been disastrous on all counts. Earlier that afternoon, Jarrett had arrived in Cologne with his producer and tour manager, Manfred Eicher, having been on the road for several days on the first leg of a European tour of improvised concerts. Jarrett had back problems and was wearing a supportive brace. He hadn't slept properly in days. And things were about to get worse.

This was to be the first time that Cologne's opera house had ever hosted a jazz concert, and a recorded one at that. When he arrived, Jarrett learned that instead of the main slot, which was in fact a traditional opera performance, he had been booked for a late-night session starting at eleven. Most disastrous of all, the piano his contract had specified was not at the venue.

Jarrett had heard that one of his contemporaries had played a specific grand piano in Germany recently, and he had asked for it to be at the venue. But when Vera Brandes, the concert promoter, put the request in to the opera house, the management response was less than ideal. Instead of ordering the actual piano, they told Brandes that they already had one at the venue. Which they did: a dilapidated, out-of-tune baby grand with faulty pedals. As he later put it, the instrument was the wrong size and sounded like an electric harpsichord: "For a renowned perfectionist such as Jarrett, who was fastidious about his pianos and possessed perfect pitch, the instrument was an abomination."

After initially refusing to play, Jarrett and Eicher succumbed to Brandes's heartfelt pleas and agreed to play the gig—after all, the sound engineers and recording equipment were already in place. A local piano tuner was brought in at the last minute to try to make the instrument more playable. Although this made some small improvements, it didn't do anything about the malfunctioning top- and bottom-end keys, or the broken pedals. As a result, Jarrett would have to play in the central keys and put a lot more effort into making the instrument audible in the massive concert hall.

To add insult to injury, the restaurant where they went for dinner served them last, and Jarrett didn't manage to eat anything before being ushered back to the opera house for the performance. Some 1,400 people turned up despite the late hour—it was a packed house. Jarrett walked onstage, sat down, and, after a momentary pause, started to play. As Brandes described it, "The minute he played the first note, everybody knew this was magic."

The Köln Concert album captures exactly how Jarrett played the instrument despite its limitations. According to Eicher: "Probably he played it the way he did because it was not a good piano. Because he could not fall in love with the sound of it, he found another way to get the most out of it." Another audience member that night shared this reflective analysis of Jarrett's performance:

> Inwardly [he was] very much in control but outwardly his playing totally freed up his passion . . . He didn't hold back at all and that immediately caught the audience . . . His left hand was almost meditative . . . just an amazing sense of rhythm . . . and with his right hand, beautiful melodies, expanding on them . . . You never did know where he would take you, like a friend, taking you on a private personal journey.

I am not the most musical of people—as my family and friends will vigorously attest—but there is even to my untutored ear something special about the journey in *The Köln Concert*. Instead of a beginning, a middle, and an end—a linear sequence, like in most rock, pop, classical, or jazz compositions—the music seems to almost unfold and bloom around different melodic fragments. The closest analogue I can think of is the rhythmic ragas played by musicians at Hindu religious festivals, where tabla players enter into an ecstatic state while playing, facing upward with their eyes rolled back so you can only see the whites.

Jarrett's playing has a similar feel, and it is easy to understand why the audience would have been entranced. Toni Morrison, the celebrated Nobel laureate and noted jazz fan, suggested that "the delight and satisfaction is not so much in the melody itself but in recognizing it when it surfaces and when it is hidden, and when it goes away completely, what is put in its place . . . all the echoes and shades and turns and pivots Jarrett plays."

Not for nothing has the album been likened to a virtuosic response to an imagined Zen stoner's query: "What's the sound of one man jamming?"

Over forty-five years later, *The Köln Concert* stands as a highlight of Keith Jarrett's career—his solo masterpiece that would turn him into one of jazz's biggest stars, and introduce him to a global audience. There was no big marketing push—the eighty-four-minute recording was made for just five hundred dollars. But the power of word of mouth would see the album become—and remain to this day—the best-selling solo jazz album of all time and also the best-selling piano album in any genre.

The Köln Concert has also become an important moment in musical history, and a potent symbol of performance in the face of pressure. Numerous commentators have written about Jarrett's achievement that night, each with their own angle and perspective. What I want to explore is the process Jarrett went through. What happened in between an exhausted musician turning up at a concert hall to be greeted with a broken piano, and the start of a performance that so wowed the audience members, and subsequently the world?

The answer is in the title of this chapter: *Mentality*. I want to explore, in his own words, Jarrett's mentality that night, and show how Upshifting not only led to his astonishing performance, but also his subsequent artistic development and musical career. More than that, I want to show how his experience carries some vital lessons for all of us seeking to perform under pressure. But before I can do that, I need to introduce you to the latest scientific thinking on how to harness stress and pressure, and how it relates to the first ingredient of *Upshift*.

FROM THREAT TO CHALLENGE

Whenever we face a stressful situation, we risk something: resources, relationships, results, or reputation. Something significant is at stake. To manage this risk, we might ask ourselves the following questions, a process psychologists term *stress appraisal*:

> *Is this situation relevant to me or not?*
>
> *Is it benign or positive?*
>
> *Could this situation cause me harm or loss?*
>
> *Am I capable of handling this situation?*

Our appraisal is made based on the things we see, hear, and feel in a given situation, on our own self-assessment, and on memories of what happened when we encountered similar situations in the past.

Of course, none of us goes about enumerating the precise nature of the stress we are going through at particular moments, and then comparing it on a spreadsheet against our resources and capabilities. For the most part, this is a deeply subjective assessment, and one that happens in an automatic and subconscious manner, although research shows that it can be made more deliberate and conscious with practice. Regardless of how the appraisal happens, we reach a conclusion as to whether the stressful situation we face poses a *threat* or a *challenge*.

A *threat* is a stress that we feel we cannot handle, because we do not feel we have the resources to meet the demands of the situation. When we interpret a situation as a threat, we ready ourselves for self-defense— the classic fight-flight-freeze response.

The threat state kick-starts a cascade of physiological responses. The heart starts beating faster. In anticipation of physical injury, blood vessels start to constrict. This is an evolutionary trick; it means we are less likely to bleed to death if we are injured. The blood flow to the brain decreases, impairing our ability to focus, and our blood pressure rises. The stress hormone cortisol is pumped from the brain into our blood-stream, raising our blood sugar levels.

This serves to give us energy for the anticipated fight or flight and also raises our sense of anxiety, often giving rise to a flood of fear-based memories. We tend to focus on the negative—the potential damage to our well-being or self-esteem—not because we believe we won't be able to succeed, but to ensure survival and to avoid danger.

This causes us distress in the moment, but if we repeatedly experience stressful situations as threats—from traffic jams to work deadlines—it can lead to these temporary bodily changes being sustained. Our blood pressure and heart rate stay dangerously high, blood vessels constrict permanently, and cortisol levels remain elevated. This leads to chronic problems over time, including increased risk of ulcers, cardiovascular disease, immune system disorders, and mental health problems.

In 1998, a large-scale and long-term US study asked thousands of adults about both their levels of stress (how much stress they experienced) and their perception of stress (how harmful stress was to their health). The participants were assessed again in the 2000s, with an astonishing conclusion. High levels of stress increased the risk of dying by 43 percent—but *only* for those people who believed stress was harmful. Those people who perceived stress as not harmful were no more likely to die. The estimates were that over the eight years since the study was undertaken, 182,000 Americans would have died because of their *belief* that stress was harmful to their health. At over twenty thousand deaths a year, that meant that negative perceptions of stress were the fifteenth leading cause of death in the United States—ahead of HIV/AIDS, homicide, and skin cancer.

What is now becoming clear, however, is that the threat state is just one end of the spectrum of responses we can have to a stressful situaation. At the other end is what psychologists term the *challenge* state, which is triggered by a potentially stressful situation that we feel we can handle, or *almost* handle, because our appraisal tells us we have the resources to meet or exceed the demands.

When we think of a situation as a challenge, we're focused on the positive: the rewards or personal growth we'll attain when we succeed. This causes what we have already discovered psychologists call *eu-stress*: forms of stressful stimulation that are associated with improved focus, accuracy, and coordination.

One of the reasons challenge and threat states have only relatively recently been differentiated from each other is that they share many obvious physiological characteristics. Both lead to raised heart rates, for example. Both lead to the triggering of cortisol. But in the challenge state, the heart rate is stronger and faster. The challenge state is one that anticipates success, and as a result there is an increase in blood flow while blood pressure actually decreases. Our hearts pump out more blood, getting more oxygen and energy around our bodies. But unlike the threat state, more of it goes to the brain, making us more alert and aware, not less. During challenge states, cortisol does not flood the body—a lower-level cortisol release is triggered compared to the threat state: instead of a tall glass of the stuff, you get a single shot, just enough to give you a blast of energy. In fact, despite superficial similarities to the physiology of the threat state, our bodies operate in a completely different way in the challenge state, more akin to how it responds to aerobic exercise (of the more enjoyable kind).

The longer-term implications of the challenge state are also radically different from those of the threat state. By analyzing people who are more likely to have a challenge response to stress, numerous long-term and in-depth studies show that there is a strong positive correlation between immune system functioning, cardiovascular health, aging, and even brain growth and size. For example, the Framingham Heart Study, one of the longest-running epidemiological studies ever conducted in the US, has shown that people who exhibit challenge responses to stress have a greater brain volume across their life spans: their brains shrink less as they age.

Most of the time, the situations we face place us on a sliding scale somewhere between an absolute threat and an absolute challenge. In other words, most situations pose some combination of threat and challenge, and it is the proportion and balance of the states that matter. And the state we are in is not fixed. Indeed, it is often quite dynamic and can change based on our perceptions, attitudes, and what we learn. In other words, we can all adapt to embrace a challenge state.

The idea of different challenge and threat responses to stress has become one of the most important in understanding performance under pressure. My research on challenge states across a whole range of high-stakes situations shows that whether you are in a challenge or a threat state is one of the best predictors of performance.

One of the most powerful examples of this is found in trauma medicine. A cohort of emergency doctors and surgeons was put through its paces in a series of resuscitation scenarios, some more critical than others. As well as gathering information before and after the exercises on the participants' states of mind, researchers also monitored the medics' cortisol and stress levels throughout the exercise. The results were remarkable: Those participants who perceived the tasks as threats had higher levels of cortisol and were more stressed. By contrast, among those who perceived the exercises as challenges, this was not the case. Because stress responses naturally influence performance, this finding was of central importance to training and supporting trauma specialists. Trauma medicine is a field I will come back to in later chapters, as it has huge relevance to performance under pressure.

As the physicians' experiences prove, the challenge state has both physiological and psychological foundations.

One of the most important insights from the stress appraisal process, and the challenge and threat model, is what we now understand about the ordering of the psychological and the physiological parts of our stress response. Specifically, our subjective assessment of whether something is a threat or a challenge *causes* physiological changes in our bodies.

This is counterintuitive: We typically think of the hardware of our brains influencing the software of our minds. Our expectation, in other words, is that the physical drives the cerebral. The reality, however, is that our stress appraisals can actually change our bodily functions. Our cognition can change our chemistry, rather than the other way around.

Neuroscientist-turned-psychologist Ian Robertson has shown that the real-time *reappraisal* of high-pressure situations can have a significant impact on the chemical makeup of our brains. As he puts it, referencing challenge and threat states and the inverted U:

> All forms of mental challenge, including moderate stress, increase noradrenaline (NA) levels in the brain. Providing noradrenaline stays near its "sweet spot," then it not only improves cognitive functioning but also helps the brain form new connections and new brain cells. . . . Stress, in moderate levels, can have positive emotional, cognitive and physical consequences, but this depends on appropriate cognitive appraisal and a resulting optimal level of arousal. Some adversity appears to be essential in life so that individuals can learn habits of managing stress, but also so that they can experience the optimisation of noradrenaline levels in the brain that properly appraised "challenge" can cause. The inverted-U-shaped function of stress—too much *or* too little can be bad for you—is related to the inverted-U-shaped function of key brain neurotransmitters such as noradrenaline. Given that noradrenaline is a key component of the "fight or flight" stress response, this is not particularly surprising.

Research by teams based at Harvard Business School has shown that the common component of all such "switching" approaches is the reframing of anxiety and distress into something positive and exciting—eustress—and focusing less on the potential for things to go wrong. This allows us to continue without the usual impacts stress has on cognition, memory, confidence, and overall performance.

Switching processes are integral to a whole range of different kinds of people doing remarkable things under pressure. They lie at the heart of each and every Upshift, and give us the scientific explanation of the click moment.

Like Sully, studies show that pilots make better use of flight data and perform safer landings when they've gone through the switching process.

Surgeons click into having better focus and fine motor skills in operations. Businesspeople do better when they're parrying with one another and click to strike better deals. Golfers make better putts, basketball players make better passes, students perform better in exams and tests, and, yes, even musicians play better.

Which takes us back to that night in Cologne in January 1975 and a hungry and exhausted Keith Jarrett.

I'M DOING THIS

When they arrived at the Cologne Opera House, the immediate response of both Jarrett and Eicher was that he should not—perhaps even could not—play. They had weighed up the situation he was facing and quite rationally decided that it just wasn't going to work. There was no perceived upside: Jarrett was hungry, in pain from his back, and facing an unplayable instrument.

This was a stress appraisal that resulted in a threat state, where he had little to gain and much to lose. But then something happened. Let's hear from Jarrett about the exact moment when he decided he was going to play and not back out:

> All I remember after the restaurant fiasco is taking a peek at the engineers sitting, waiting with their equipment. They had everything ready. And I started thinking, "I'm going to do this." I remember putting my fist up in the air on the way out [from] backstage. I just looked at Manfred and [said], "Power!"

This was Jarrett's click moment: looking at the engineers waiting with their equipment and deciding that he was "doing this." He had just Upshifted into a challenge state. His use of the elevated-fist gesture is also telling, a defiant symbol of resistance and strength against adversity.

According to stress psychologist Alia Crum, "Making a deliberate shift in mindset when you're feeling stressed is even more empowering than having an automatically positive view." Imagine, for example, that you are about to give a talk onstage. Instead of thinking you might trip up the stairs, get tongue-tied, or lose out on laughs when it's time to deliver the punch line, you might focus on your strengths: how excited you are to take center stage and how well prepared you are for the performance you are about to give. Such stress reappraisal mechanisms are well understood by sports psychologists, who use a variety of techniques to help top athletes to stop fighting the "butterflies" in their belly and instead *get them to fly in formation.*

The very next thing Jarrett felt onstage was a wave of buoyant solidarity: "Then something interesting happened. It just seemed like everybody in the audience was there for a tremendous experience, and that made my job easy." This is not an uncommon experience among live performers—being lifted by the communal feeling of being together to experience something new and different in a particular moment. This will of course be familiar to those of a religious or spiritual leaning. Indeed, Jarrett himself says that this unspoken audience engagement is "close to what I think communion ought to be." More than this, however, the positive framing of the audience's expectation of a "tremendous experience" as making his job easy is another indication that Jarrett had shifted from a threat to a challenge state.

The mindset shifts are necessary but not sufficient: he still had to play the broken instrument. "What happened with this piano was that *I was forced to play in what was—at the time—a new way.* Somehow, I felt I had to bring out whatever qualities this instrument had. And that was it. My sense was, 'I have to do this. I'm doing it. I don't care what the fuck the piano sounds like. I'm doing it.' And I did" [emphasis added]. Again, the sense of rising up to meet the challenge is palpable.

Despite his subsequent distancing from *The Köln Concert* recording—he once said that he wished every one of the 3.5 million copies could be stomped into the ground—the experience itself clearly set a precedent for his career of improvised concerts. Interviewed in 2017, over forty years later, Jarrett described how he made a habit of sitting down in front of his audiences and moving his hands to keys he had no intention of playing.

> So, when I find a piano that has this "imperfect" character, it's actually much more to deal with—and I mean that in a good sense—than a "perfect" piano. You're hearing me discover which notes on the keyboard will do this . . . I'm learning what part of the keyboard is acting a certain way.

Remember back in Cologne Jarrett had to stay within the central keys of the piano and use his left and right hands in ways that made up for the lack of tone, timbre, or sustain from the pedals. To my mind, playing the middle of the broken piano is analogous to Jarrett reaching the peak performance zone on the inverted U of the Yerkes-Dodson curve. And once there, he had to do everything possible to stay in it. Again in his words, "It takes everything to do it. It takes real time, no editing, it takes your nervous system to be on alert for every possible thing."

This is the space Jarrett would go on to master. As described by one music critic: "Absolute lucidity in this transient sound, created without the possibility of replication, has become his trademark." From this click moment, Jarrett would go on to make a habit of reframing stress and pressure: "I just learned [and] I . . . successfully reversed some kinds of thinking that aren't helpful—for example, that what is desirable is to eliminate stress from your life. If you eliminate stress from your life you eliminate life from your life."

Keith Jarrett's performance in Cologne is a stellar example of "doing this." Instead of "Keep Calm and Carry On," as the ubiquitous UK poster has it, Jarrett's mantra would have been "Get Amped Up and Do the Unexpected. And Then Keep Doing It."

Moreover, like Sully's 208-second version, the story of the Cologne concert provides us with the critical ingredients of Upshifting in a nutshell: mentality, originality, and purpose fusing together in sixty-six remarkable minutes, captured for posterity. Go and listen.

So how did Jarrett learn to Keep Doing It? And how can we do the same?

THE STRESS MINDSET AND THE POWER OF ADVERSITY

Keith Jarrett's experience that night in Cologne, together with the idea of challenge states and stress reappraisals, explains a first vital step of Upshifting: the empowering switch between threat and challenge states.

It also gives a new and powerful way to understand some of the other vivid Upshifts I have already shared, including my experience on the barge in Sri Lanka.

You will have noticed that these examples—Sully heading for the Hudson, Jarrett sitting down at the broken piano—are highly situational and unplanned. But these moments can have a lasting impact on behavior. Think of how Jarrett turned a serendipitous moment into a repeated habit.

Just as Jarrett developed the skill of feeling for the imperfections of each piano, and used this to catalyze better performance, it is possible to use serendipitous Upshifts as the starting points for more conscious and deliberate practice. We can all move from one-off Upshifts to establishing a *mindset* that fosters the likelihood of Upshifts.

Such a mindset was described by Stanford University scholar Carol Dweck in her popular book of that name (*Mindset*). She found that some of us have particular "ways of thinking about a problem that's slightly too hard for you to solve." Growth mindsets involve operating just outside of our comfort zone when we tell ourselves "we just haven't solved it *yet*." Dweck argues that growth mindsets are key to thriving in challenging times, and moreover are central to all kinds of deliberate practice.

Psychologists specializing in understanding stress have taken her ideas and used them to develop a theory known as *stress mindsets*. While stress appraisals provide an immediate assessment of our resources to cope with the demands of a particular situation, our stress mindset assesses the situation both in the short term and the long term in light of our belief about the nature of stress, and our own past performance in relation to stress.

Some of us are predisposed to the mindset that stress carries *enhancing* consequences for performance, productivity, learning, and growth. It has been shown that with this mindset, we are likely to have more moderate physiological responses to stress: we will have lower levels of cortisol release. It also has psychological implications: for example, we will be more likely to deal with stress in a socialized way, being more willing to seek out support and feedback. (Think back to Jarrett feeling the crowds' tremendous expectations buoying him upward.)

For example, you might have a mindset that "stress is enhancing," but also frequently find yourself in high states of anxiety about high-profile public-speaking engagements. This motivates you to practice more, do test runs, and draft in colleagues and friends to help. If things go well, this might lead you to have a more positive assessment of your next such engagement.

Others among us might be more of the mindset that stress has *debilitating* consequences for performance, development, and health. If you were preparing for a talk, you might avoid doing any preparation until the very last minute, feel overly anxious, not communicate your fears to anyone, and ultimately decide to pull out on the day, citing unforeseen health problems.

I have done both of these things, incidentally, in the context of public speaking. When I first started my career, I was so nervous about speaking up in meetings that I could hear my heartbeat—"Thump. Thump. Thump."—in my eardrums. And this was just in a room with usually eight people informally sharing ideas. I was so shy and lacking in self-belief that I'd freeze. Sometimes I froze so badly before talks I had to ask my boss to give the talk. Once I even faked having lost my voice because I was so nervous about giving a presentation. This was in my early twenties. By my thirties, thanks to hard work and practice, I was regularly giving keynotes to conferences of hundreds of people.

This goes to show that our stress mindsets are not fixed—our disposition does not need to be our destiny. It won't come as a surprise that our stress mindset plays a critical role in determining how we deal with different situations. Because mindsets are anchored in beliefs, they are not situation specific but are more reflections of our current philosophy and experiences of life. They both shape and can in turn be shaped by the stress appraisal processes I described earlier in this chapter.

Looking at the research on stress mindsets, there are three big messages for habitual Upshifters. First, it may not be surprising to learn that the "stress is debilitating" mindset is much more common, regardless of gender, age, and ethnic background. Most of us believe that stress is bad for us, without qualification.

Second, despite this, we can change our attitude and mentality toward stress. For a whole range of problems and issues, people have been shown to be able to switch to a "stress is enhancing" outlook. We can do this through a variety of means: we can try to be better informed about the stresses we face, we can get feedback about how we are doing, and we can seek out mentors who have done it before. All of these things help to reframe how we see potential threats. My own experience of public speaking is testament to this.

Third, if we are able to adopt a "stress is enhancing" mentality, we are more likely to perform better in a whole variety of situations. This is not just proved by positive psychology, but by our biochemistry and physiology: from job interviews to group problem-solving, experiments show that the peak performance is linked to cortisol levels that are neither too high or too low, but on the sweet spot of the Upshift curve. And it goes beyond performance: if we can see stress as enhancing, we are even likely to live longer than if we are stuck in seeing it as a threat.

Keith Jarrett's description—"if you eliminate stress from your life, you eliminate life from your life"—is a beautiful, pithy summary of the science of the "stress is enhancing" mindset.

I want to draw on findings from very different groups whose experiences illustrate these three insights: university students and Olympic athletes. Research on how the stress mindset operates for both of these groups, working in very different but challenging contexts, illustrates the power of the concept.

Despite their apparent privilege, young people studying at university are one of the most stressed-out groups in society. According to a number of indicators, including mental health issues, nervous breakdowns, and suicide rates, they are among the most vulnerable groups in the eighteen-to-twenty-five-year age range in the developed world.

A study of Australian students at leading national universities sought to assess whether these stress levels could be addressed through training and learning. Having identified the existing stress mindsets of hundreds of students, they then put them through a range of meditation and visualization processes that encouraged a "stress is enhancing" mentality. Some of them were directly linked to the student experience (taking exams, presenting in front of peers) while others were more general. In all of them, the students were asked to imagine peak performance: overcoming the threat and mentally rehearsing what success felt like.

The theory behind this is that mental imagery is an effective means of rehearsing and instilling psychological states that, in turn, influence behavior. Much like learning by observation or through simulations, visual mental rehearsal is seen as stimulating neural networks related to what is being imagined, making them accessible for when we find ourselves in similar situations in real life.

After the courses were completed, the evaluation found two significant changes. First, a significant proportion of the students had switched to a "stress is enhancing" mentality. Second, their overall levels of distress and anxiety had lowered, and, as a result, academic performance and general well-being had been positively impacted for the duration of the study. And the positive gains were most pronounced for those students who had started the process with a "stress is debilitating" mentality.

We can get a longer-term perspective on the power of stress mentalities thanks to the UK's high-performance sports agency, UK Sport. Sports scientists and psychologists have long known about the vital importance of overcoming adversity for athletic performance. What UK Sport did was explore what role this plays in separating the very best from the rest. They selected thirty-two high-profile elite athletes who had represented the UK internationally, with half being super-elite—those who won multiple medals at Olympic or world athletic events. And they went about excavating huge amounts of data about each athlete: everything from how many hours they had trained since childhood, the time spent playing alternative sports, and even the size of their hometowns. They combined all this with in-depth interviews with the athletes and their family members, coaches, and peers.

The volume of data is staggering—the interview transcript material alone exceeds some 2.4 million words on almost ten thousand pages (almost three times as long as the complete works of William Shakespeare). The data had to be analyzed using artificial intelligence tools that were trained in pattern recognition and could identify correlations and the strength of relationships between many different variables.

While the studies are a treasure trove for anyone interested in performance, sports, and psychology, the most fascinating parts for us are the factors that separate the super-elite and elite performers. Seven ingredients were identified as the ones with the most robust evidence base. Of these, almost half related directly to how the super-elite dealt with adversity, stress, and pressure.

All of the super-elite athletes experienced early-life trauma or adversity that was closely connected to a positive sporting experience. This set a template for "post-traumatic growth" that would shape their subsequent professional development.

They experienced greater total numbers of substantial year-to-year performance setbacks; in particular, substantial setbacks after having performed "near career peak performance" level.

Last, the super-elite athletes experienced heightened levels of pressure and anxiety in high-level championships, but they also had a mentality of what the researchers termed *counterphobia*. This was defined as a response to stress that, instead of fleeing the source in the manner of a phobia, actively seeks it out, in the hope of overcoming it. This is what we earlier called a *challenge mindset*. All of the super-elite were drawn to, purposefully "tackled," and in some way even enjoyed, high-pressure situations—the very competitive situations they then thrived in. The super-elites didn't avoid pressure or anxiety: they had developed ways of running toward it.

The overall conclusion was striking: Athletes who experienced adversity while they were growing up, and who coped with setbacks and stresses, all things being equal, developed a stress mindset that gave them a competitive advantage over those who did not experience adversity. Taken as a whole, how super-elite athletes deal with stress and adversity was the single biggest contribution to their success.

This is not to say that adversity should be seen as a good thing, or that athletes should be deliberately exposed to extreme stresses and strains to make them reach higher levels of performance. Such a conclusion would be unethical and potentially abusive and harmful—as recent evidence in certain Olympic training regimes, notably in women's gymnastics, has revealed.

But there are some sound implications for athletes' development, and they apply much more widely: "It is important to encourage [them] to actively engage with challenging situations that present opportunities for them to raise their performance level." This means "expos[ing] them to appropriate and progressively more demanding stressors." Like

Keith Jarrett feeling for the imperfections on a new piano and trusting that he will find music, we all need to be aware of how our stress mindset shapes our response to pressure, and learn how we can switch it to an enhancing perspective.

THE MASTER SWITCH OF LIFE

How long can you hold your breath? For most of us, this would be in the order of a couple of minutes at most. When we hold our breath, we feel a strong instinct to inhale that is triggered not by a lack of oxygen but by the elevated levels of carbon dioxide that we are not exhaling from our body.

But some people specialize in overcoming this. The world record for holding one's breath is a staggering twenty-four minutes and thirty-seven seconds, set in 2021 by a freediver, one of those extreme athletes who specialize in diving to astonishing depths without equipment, and with only a single breath of air. Imagine holding your breath for an entire episode of *Friends*. Plus two minutes.

On my own introduction to freediving, in the rather unglamorous location of a UK public swimming pool, I held my breath until my diaphragm contracted and hit what scientists call the "struggle phase," and then let out big lungfuls of air in relief. I lasted about twenty-five seconds the first time, and improved ever so slightly over the course. What made for this difference?

Much of the research on freediving focuses on the physiology of the divers. Their remarkable achievements are thanks to their exploitation of the mammalian dive reflex—also known as the "master switch of life." This is a whole set of reflexes that kicks in whenever we enter the water. It changes the functioning of our brains and lungs, and also protects us from the underwater pressure of deep water. It is thanks to the master switch that divers can survive pressures in the water that would be dangerous on land—and it is this that freedivers exploit in their astonishing efforts to reach ever-greater depths. In the 1970s, the freediving record was nearly 100 meters; today it stands at 214 meters. When I spoke to freedivers themselves during my own amateur efforts, they suggested that the largest challenge in the sport—and the main reason for this remarkable progress—is not physical but psychological prowess.

This is borne out by the insights of competitive freedivers, research among whom has shown that psychology may indeed be a greater determinant of success than physical traits. Successful divers are especially adept at dealing with the "struggle phase." In this phase, the techniques used include a whole range of stress reappraisal techniques, such as visualizations akin to the ones used by our stressed-out students earlier, counterphobic techniques used by Olympic athletes, and more.

The key to all these situations is to anticipate and reframe stress that will be inevitably experienced on each dive. One particular diver, Croatian Katarina Linczenyiova, describes her technique as follows:

> During visualisation, I imagine my whole dive, every second and every step I do. In real life, one dive to 90m . . . takes about three minutes. In visualisation, it can take an hour. I can do that dive 100 times in my mind and prepare myself for different scenarios. Through it, I can find out what stresses me and then work out how to handle that emotion.

There is a growing body of research and practice that resonates with these experiences of freedivers, which has the wonderful label of "stress inoculation." Like with biomedical vaccines, stress inoculation is based on the principle that exposure to adversity in moderation can help individuals to develop appropriate mindsets to cope with future pressure situations. These techniques can include:

- Stress training and testing, individually and in groups
- Systemic exposure to critical incidents through simulations and scenarios
- Dealing with challenges of evaluation and judgment in a controlled and safe environment
- Using setbacks and failures as a focus for learning and collective reflection

Recent research from human evolutionary biology suggests we are primed for "stress inoculation" from an early age through play-fighting. Typically, this involves simulating some more serious competitive behavior, but doing so with some form of cooperation that makes the interaction safer and more enjoyable. Research on children across different cultures suggests that the most important aspect of play-fighting for individual and social development is that it forces participants to carefully monitor their actions, as well as those of their partners, and to ensure a degree of reciprocity.

The shared creation and testing of boundaries—of discomfort, pain, social norms, and so on—is essential to developing a more positive approach to future stresses. Play-fighting is more than just fun: It amounts to training for the unexpected, which helps to develop "stress as enhancing" mindsets and pressure adaptability in adolescents and adults. Play when young literally trains us for the unexpected when older.

This makes me think back to the experiences I opened the book with, and how in different ways I played my way through the worst of the civil war in which I grew up. I remember listening with all the grownups in my family to a radio broadcast that was reporting on what was happening across the country. The broadcaster was talking about people being squeezed into used car tires in twos and threes and set on fire while still alive. I was playing with some small seashells under the table on which the radio sat. My uncle, who was the tallest and strongest person I knew, was standing by the radio, and I heard him starting to cry. Then my grandmother turned the radio down, telling the other grown-ups to get me and my brother out of the room. I remember carrying on playing with the shells afterward, outside the room. I distinctly remember thinking, as long as I have the shells, and I am playing with them, I'll be okay.

In the weeks, months, and years that followed, until my family was lucky enough to escape the war, playing games was like a protective force field around me. It was my own fantastical escape from reality. Everything could be turned into a game if you tried hard enough. Escaping from your house at nighttime before the bad men turned up to set fire to the town? The "Keep quiet" game. Having to leave behind your beloved family pet because he might bark? The "What will my dog be doing today?" game. Walking across stinking mudflats for hours on end so you and your family don't leave footprints? The "How sticky can your shoes get without getting lost in the mud?" game. Traveling on a barge with armed soldiers, and deciding to make friends with them, despite the terrified protests of the grown-ups? The "How long will an army man let me hold his gun?" game. I literally play-fought my way out of a horrific situation. Playing games helped make sense of and bring some order to it all. I'm not for a moment suggesting that this coping mechanism can or should help with everyone's traumatic experiences, but this was my own version of the Olympic athletes' "post-traumatic growth": using the trauma of the war to catalyze my own imagination, to escape from reality into a place where I could play, and therefore feel safe.

Perhaps the most important message about Upshifting that I want to reiterate in this chapter is that stress appraisals and reappraisals, stress-enhancing mindsets, and stress inoculation approaches are not inherited, nor are they fixed. Capitalizing on a positive stress mentality is something to be learned, practiced, and improved. Keith Jarrett's whole tour prior to Cologne was based on improvised pieces—each concert was unique. He was, in many ways, in training for the unexpected. It was just that the conditions in Cologne proved the trigger for unusual brilliance. Seen in this light, the story is not just one of performance in the face of pressure. It also echoes an old adage: Luck is what happens when preparation meets opportunity. And this preparation is not just for epic challenges such as playing broken pianos in opera houses, winning gold medals at the Olympics, or diving to record depths without equipment. Doctors and surgeons do it on a daily basis. Pilots do it. Businesspeople and office workers do it. Stressed-out students do it. Even play-fighting kids do it.

Two-time Super Bowl champion quarterback Eli Manning was once asked whether handling pressure was in his genes. His reply is exemplary of the Upshifting mentality: "No, it's because I practiced the plays thousands of times in all different scenarios . . . In pressure situations, I never think of failing; I think of all the past times I have succeeded."

2

ORIGINALITY UNDER PRESSURE

GREAT MINDS THINK DIFFERENTLY

THE MOMENT OF TRUTH

The biggest, highest-stake game in Las Vegas runs just a handful of times a year. It costs billions of dollars, and has no croupiers, cards, or dice—but the lights are spectacular.

Watch the skies from a distance on game nights, and it is as if the city has become the launchpad for a meteor shower, stars catapulting across the night sky in rapid, unearthly lines. Keep watching and you will see each of the stars return, one by one, until the skyline is dominated once more by the hard glitter of the city of lost wages.

These aren't shooting stars, of course, and there are no gamblers to be found at this game—at least not officially. This is Nellis Air Force Base, located on the northeastern edge of the city, and we are witnessing what is perhaps the most serious game in the world: the US Air Force's Red Flag air-to-air combat training program. Each Red Flag mission involves protracted aerial engagement between fighter planes belonging to the "red" forces of an imaginary advancing enemy and the "blue" forces, made up of both US and NATO partner countries. The simulations include attacks and strikes, reconnaissance efforts, air lifts, and refueling.

The scale is nothing short of breathtaking. In a single year of exercises, Red Flag missions involve five hundred aircraft flying more than twenty thousand sorties, involving five thousand air crew members and fourteen thousand support and maintenance personnel. This all happens across a cordoned-off area in the Nevada desert that is half the size of Switzerland.

The exercises that unfold take a year to prepare, last several weeks, both day and night, and have been described by one senior military leader from Great Britain as "the best real-world training without actually going to war." At the heart of it all is a computerized system of digital targets and sensors that means not a single actual weapon is fired. This is laser tag as you have never seen it before.

While the Red Flag exercise is obviously noteworthy in its own right, my reason for introducing it is because of what it conveys about the hinterland of perhaps the most famous former participant. More than thirty years before he became a national hero for his actions in landing a downed airplane on the Hudson River, Captain Chesley "Sully" Sullenberger was a Red Flag mission leader at Nellis.

As we saw in the introduction, Sully's experience aligns with all three principles of *Upshift*. But it is original thinking under pressure that is the focus of this chapter. What Captain Sully and his crew faced in that split second is what surgeons call the "moment of truth." On the operating table this is the moment when an aneurysm ruptures, a brain swells, or a tumor bleeds profusely: "a moment during impending catastrophe when all that stands between life and death is the surgeon's experience, knowledge, and decision-making capability." When facing such an acute operational catastrophe, in the face of intense pressure, many surgeons make the decision to stick to existing, known approaches. Some do not. Apply the same principles to Flight 1549: Sully is clearly in the latter category. How exactly do you solve a problem you have never seen before, in such pressured circumstances? What I want to do in this chapter is explore why and how he got there, and how we can too.

Prior to that fateful morning in New York, Chesley Sullenberger's flying record was exemplary. Having learned to fly at the age of sixteen, he clocked up a record number of hours in the air during his time at the US Air Force Academy, won Outstanding Cadet in Airmanship in his year of qualification, and was a flight instructor by the time he was twenty. After the military, as a civilian pilot Sully was an instructor, a safety chairman for the US Airline Pilots Association, and a long-standing accident investigator for both the National Transportation Safety Board and the US Air Force.

Sully had a firm grasp of everything from high-altitude aerodynamics and physiology to airplane mechanics and navigation functions. And he was steeped in detailed knowledge of major flight accidents: where they occurred; why and what was learned from them; and how that knowledge then informed everything from plane designs to policies, procedures, and training. When he decided to land the plane on the river, Sully knew exactly what he had to do, and what he should not do. He had flown gliders in the military, which gave him the operational skills to maneuver the engineless plane. He had published articles on crew management, and knew how to get the whole team working together. And he knew of an airplane that had ditched in the Indian Ocean and shattered, killing everyone on board, and understood precisely what this meant for the landing approach he needed. The choices that he faced on that cold January morning would test his expertise to its limit.

What Sully did—like Keith Jarrett, in the previous chapter—was not to follow convention, but to break out, consider all possible options, and try something novel to address a problem he had never seen before. And get it right the first time.

Sully led the missions in the Red Flag exercises out of Nellis for several years. He described his time flying fighter planes vividly: "like driving a Formula One racer on steroids and in three dimensions." More than the thrill of flying, he also clearly valued the problem-solving rigor and discipline.

After each simulated mission Sully flew, he was involved in brutally honest discussions about what worked, what didn't, and how things could be improved. These were simulations; it was war-gaming. But the whole point was to play and then to learn. He carried this rigor with him for years: "Anyone who has been a military aviator brings to the table a certain discipline, a certain diligence. It's an attitude with which we approach the job, it's a profession, it's a calling. And it's something that if you answer that calling, it leads one to be a continuous lifelong learner, to constantly be striving for excellence to try to always make the next flight better than the previous one."

Thanks to Red Flag simulations, and his military training, Sully was well prepared to handle "moments of truth."

Like the US Navy's more famous TOPGUN training program, Red Flag has its origins in the high-profile failures of US aerial combat during the Vietnam War.

Despite facing what was on paper a much weaker and less well-equipped foe in the Vietcong, US forces suffered terrible aerial combat losses. The success of fighter pilots is measured by the "exchange ratio" of enemy aircraft downed compared to US aircraft. This number was perilously low, at some points even dropping below one: meaning the Vietcong were downing more US aircraft than the other way around. Although the diagnoses differed across the different parts of the military, one conclusion was clear and consistent: US fighter pilots needed better training for combat.

Both the TOPGUN and Red Flag programs were designed to get back to the fundamentals of air-combat maneuvering in the wake of large-scale failure in Vietnam. TOPGUN was based on training teachers who could then instruct their squadrons about warfare tactics (which was the basic premise of the 1986 movie *Top Gun*).

The Red Flag exercises were designed to give pilots their first "combat" exposure in a realistic but controlled setting. Noting that pilots face the steepest learning curve during their first ten combat missions—when they are most likely to be downed—the US Air Force decided to hold regular simulations of those most dangerous missions in large-scale repeated exercises.

The whole point of these exercises wasn't to act like Maverick, strike out alone, and win—it was to practice, cooperate, and learn.

In a roundabout way, Sully himself anticipated his own remarkable achievements in an article on major airplane accidents, which he coauthored with NASA scientists and published ten years before the Hudson landing. In it, he and his colleagues found that one of the most common causes of disasters was "plan continuation error." Specifically, in more than 75 percent of all disasters, pilots often failed to consider all the available options and persisted with a narrowly defined approach, regardless of the novelty of the challenge they faced. Plan continuation error is attributed to organizational and social factors, which create conflicts between the pilots and those around them. Flick back to the transcript of Flight 1549, and the sense of confusion among the traffic controllers at Sully's statement ("We're gonna be in the Hudson") is clear.

Captain Sully's Upshift is an extraordinary example of original thinking under pressure. In the chapter ahead I want to really dig into what it is, how it works, and how we can all get better at it. But the big message I want to convey right at the start is that, despite being shrouded in myths and misapprehensions, original thinking is a practice and a discipline that every one of us can learn. Don't believe me? Let me prove it using a paper clip.

J. P. GUILFORD AND THE DISCOVERY OF DIVERGENT THINKING

How many different ways can you use a paper clip? Think about it for a few minutes. Make a list, and write the answers down if you like.

Some of your answers will be pretty straightforward. You can use the existing function and design of a paper clip and use it to "clip" things together: To clip cash, flower stalks, or other things; to securely close food bags; as a cheap bookmark; to hang Christmas baubles on the tree. So well-established is the paper clip's basic functionality that few of us are consciously aware that this simple physical device has become universally used as the symbol for email attachments.

Other answers probably involve repurposing the paper clip as a tool. You can unwind it to create a handy cleaning tool that can get under nails, into clogged-up saltshakers, or drains, or to be used as a device to reset electronics, pick padlocks, or move items around under a microscope. You can use it as a precision scratching device when you have an itchy limb in a cast after an injury, to puncture balloons after a kids' party, or to open a tube of paint.

More original uses might include using a paper clip as a device for making marks: to write messages using ink, dabbed in paint to create art forms, and so on. Paper clips can be used to make jewelry, clothing, machines, and furniture. They can be placed en masse into a can to make a percussion instrument. They can be melted down to make new metal objects. They can even be used as a coded communication tool during covert wartime operations (just look up "Operation Paperclip" if you don't believe me).

By creating your list, you have just performed what's called the Alternative Uses Test—one of the most established means of assessing original thinking. Most of us will begin our responses to the test with variations of the established uses of a paper clip. However, after a number of more-or-less conventional ideas—the magic number for most people, according to averages taken across a number of synthesis studies, is nine—the ideas will start to become more original and depart from convention.

We will return back to some of the insights that we have thanks to the Alternative Uses Test and its variants later on in this chapter, but what I want to do now is explain its origins. The answer takes us back to fighter pilots.

You may be surprised to learn that creativity as a subject of scientific study is relatively new. Prior to 1950, creativity was seen as fuzzy and subjective, and systematic research was minimal. It focused on famous artistic or scientific geniuses—think Mozart, Austen, and Curie—and tended to look at the inherent traits that underpinned their creativity. The study of such an imprecise topic was seen as hard to reconcile with the then-dominant Pavlovian approach of stimulus and response to understanding behavior, and IQ-based assessments of intelligence.

After 1950, there was an explosion of research programs, corporate laboratories, and professorships. Creativity was seen not as a trait, but as the *process of original thinking*. Moreover, creativity was not seen as confined to a few special individuals, but identifiable—and therefore attainable—among the population at large, and across a very wide range of problems or tasks. Creativity was seen as within everyone's grasp; it had become democratized both as a subject and as an act.

The reason for this shift is largely attributable to an intellectual giant in the field of psychology, J. P. Guilford, and his work during and after the Second World War. Guilford was chief psychologist for the US Air Force, and oversaw a transformation in the way pilots were selected, trained, and developed over the course of their careers that would result in the approaches that would so benefit Captain Sullenberger over the course of his career, not least on Flight 1549.

Before the Second World War, pilots had been selected on the basis of physical tests and IQ measures. But as the war escalated, flight technology advanced and battles became more complex. The air force top brass were puzzled that IQ seemed to be no indicator of whether pilots would survive or not. A very intelligent pilot confronted by an emergency situation would go "by the book," even when the book had no solution for the problem.

By contrast, less-clever pilots might undertake a daring but risky maneuver and survive. Guilford, a leading academic of his day, was made chief of psychological research in the air force, and was tasked with developing creativity tests for use in the selection of pilots, bombardiers, and navigators. The US Air Force asked him to find out why intelligent and capable aviators weren't equipped to deal with the unanticipated problems that emerged when their planes were hit and damaged in combat.

When initially tasked with the role, Guilford developed a system that involved tests, a grading system, and personal interviews. Also assigned to the selection process by the air force was a retired flight officer without any academic training. Guilford did not have much faith in the officer's experience. It turned out that Guilford and the pilot chose different candidates. After a while, their work was evaluated and, surprisingly, more of the pilots chosen by Guilford were shot down and killed than those selected by the retired officer.

Guilford later confessed to being so depressed about sending so many pilots to their deaths that he considered suicide. Instead of this course of action, he decided to find out why the candidates chosen by the retired pilot had fared so much better than those he had selected:

The old pilot said that he had asked one question to all the would-be pilots: "What would you do if your plane was shot at by German anti-aircraft when you were flying over Germany?" He ruled out everyone who answered, "I'd fly higher." Those who answered, "I don't know—maybe I'd dive" or "I'd zig-zag" or "I'd roll and try to avoid the gunfire by turning" all gave the wrong answer according to the rule book. The retired pilot, however, chose his candidates from the group that answered incorrectly. The soldiers who followed the manual were also very predictable . . . The problem was that even the Germans knew that you should fly higher when under fire and their fighter planes therefore lay in wait above the clouds, ready to shoot down the American pilots. In other words, it was the creative pilots who survived more often than those who may have been more intelligent, but who stuck by the rules.

Guilford used this experience to develop tests that would assess the ability to think differently, unexpectedly, and creatively under pressure to select those pilots who were most likely to be able to improvise and come up with unexpected solutions. He would later describe the approach of the successful fighter pilots as "divergent thinking," in contrast to the "convergent thinking" of the unsuccessful pilots.

Convergent thinking is the process of figuring out the most effective answer to a problem. It generally involves using previously identified techniques and applying them to new problems. A whole range of tests—such as multiple choice, standardized tests, quizzes, spelling tests, and the like—require convergent thinking, because only one answer can be correct. One of the biggest criticisms today of IQ tests—which prior to Guilford had dominated pilot assessments—is that they focus almost entirely on testing for convergent thinking skills. This explains why plan continuation error is so prevalent.

Divergent thinking, in contrast, is a process of thinking that explores multiple possible solutions in order to generate creative ideas. It generally involves figuring out new procedures to solve a problem despite existing solutions.

Guilford developed a whole battery of structured tests and assessments for convergent thinking. But his most famous—and still widely used—test was for divergent thinking: the Alternative Uses Test I introduced with the paper clip. The test was specifically designed by Guilford to overcome the shortcomings of the tests that he felt had failed the fighter pilots, all of which examined whether or not their answers were correct.

Think back to your own list from earlier. Try to fight the compulsion to be judgmental. The list doesn't tell a story about whether you were right or wrong; instead it tells the story of how you think about problems. If you use it right, it can be a tool for improving, not proving, your creativity.

If you want to assess your creative thinking, these are the criteria that Guilford and his colleagues developed:

> Fluency: *This was the total number of alternate paper clip uses you could think of. The greater the number of options you generate, the greater the chance of coming up with something original and useful.*
>
> Flexibility: *The different categories of ideas that you generated. This would assess how well you could see situations from different perspectives, and reframe the problems being dealt with.*
>
> Originality: *How far your ideas were from the norm. The "norm" here being using paper clips to hold things together. The greater the number of unfamiliar, unusual, or novel ideas, the more likely you would see something genuinely original.*
>
> Elaboration: *The level of detail behind your ideas. The more thought through and considered, the more likely it was to be brought to life and survive the "encounter with reality."*
>
> Resistance to "premature closure": *Your ability to keep an open mind and delay closure of possible options long enough to make the mental leap into originality.*

Each of these criteria proved especially relevant to life-or-death decisions that fighter pilots would routinely face.

Research led by Guilford's colleagues using variations of the Alternative Uses Test showed that "ace" fighter pilots not only came up with more total responses (fluency), but also more unusual content categories (flexibility), and were more imaginative in their thinking (originality). Moreover, their divergent thinking was characterized by a "singleness of purpose and intensity of effort" (elaboration and resistance to premature closure). In other words, not only did they do original things, but they did so rigorously. I think it is possible to see in here the origins of the practice that so benefited Captain Sullenberger (and his crew and passengers).

So how can we all practice divergent thinking in moments of truth? Our appraisal of the problem we are facing is crucial. As we learned in the last chapter, if we assess the stressful situations we face as challenges as opposed to threats, we are more likely to be able to perform effectively. Research has shown a further dimension to this: Divergent thinking is also more likely when we appraise a situation as a challenge. Being in a challenge state means we are more willing to explore possibilities.

Framing stress and pressure in a positive way can broaden and enhance our thought repertoires, produce consistently novel ideas, and permanently expand our scope for original actions.

In the situation Captain Sully found himself, the obvious answer was to land at one of the alternative runways. Most of us would have narrowed our choices down to the convergent preexisting solution, especially given the instructions from the authorities. But Sully did not do that. The bird strike was an accident, but Captain Sullenberger's decision was no accident. As he put it himself, he had a lifetime of training—through flying, simulations, and major-incident analyses—and he was able to bring all these to bear that day.

Think back to how Sully described overcoming the extreme stress of the bird strike, forcing himself out of the threat state and into a challenge state. Once there, he rapidly explored the different possibilities that were open to him, and discounted those that were not. In such exploration processes, our individual preferences and background play a significant role. Some of us have a high desire for closure in certain stress situations; we literally shut down alternatives quicker, and move toward the seeming "right answer." Some of us, however, have—either through nature or nurture—a low desire for closure, and are more able to suspend our desire for certainty while working through alternative, novel options.

And again, Sully would have known this all too well, thanks to his work in the 1990s with NASA. Premature closure to divergence and plan continuation are, after all, different ways of describing the same decision-making bias.

These two linked capabilities that underpin divergent thinking—willingness to explore and low desire for closure—have been identified again and again in many different fields. One of the most famous post–Second World War studies on creativity looked at some of the leading architects of the day and attempted to understand what separated the most original from the least original. Two factors stood out, which will remind you of a certain captain's actions.

Like Sully, the most creative architects were better at playing with knowledge: "a knowledgeable person . . . is not merely one who has an accumulation of facts, but rather one who has the capacity to have sport with what [they] know, giving creative rein to [their] fancy." And they also actively practiced not prejudging new ideas, and "deferring decisions."

The Alternative Uses Test and related approaches to assessing divergent thinking are extremely robust from a scientific perspective: They help to explain and predict creativity in moments of truth with remarkable success. They are as rigorous as IQ tests, which so troubled Guilford during the war, and time has proved them more reliable and valid than almost all other behavioral psychology tests. This simple assessment gives truth to Albert Einstein's famous quip—which Sully demonstrated so well on that cold January morning: "imagination is more important than knowledge."

Although as Einstein would probably have been at pains to point out, imagination is not fixed. The results of divergent thinking tests are not rigid (unlike IQ tests). They offer an estimate of our creative potential—and this potential is not fixed, but evolves over time with practice. No matter how good we are at divergent thinking to begin with, we can all get better at it.

Let me pause here and take stock. Not only had the Upshifting phenomenon been observed in the past, but J. P. Guilford's firsthand analysis of original thinking under pressure—the divergent approaches of Upshifters operating in the high-stress context of military aerial battles, the forerunners of Captain Sully—fundamentally transformed our very understanding of what creativity is and how it can be strengthened.

In all my work on *Upshift*, this is perhaps the finding that still astonishes me the most. I had the strongest sense that *Upshift* was not something I was crafting, but something I was discovering.

After the war, Guilford was made president of the American Psychological Association and gave an inaugural keynote in 1950 titled "Creativity," in which he set out some of the lessons from the US Air Force. He argued for more research on creativity—which he defined as the capacity for divergent thinking—as fundamental for national and global progress in the postwar era, especially in light of the looming Cold War face-off. The speech was followed by an explosion in studies and laboratories dedicated to the topic, so much so that Guilford is today widely acknowledged as the "father of modern creativity." And perhaps the most important message from his work is that every single one of us has the potential to be creative and can work to get better at it.

That this might seem surprising to those of us who do not see ourselves as creative is in part because of one of the most startling cases of misinterpretation and misunderstanding of science in recent years. To understand it, we have to move from psychology to neuroscience: from the mind to the brain.

THE WINDING ROAD TO THE CREATIVE BRAIN

War casts a long shadow over many lives, but amidst the darkness, there are also bright spots. In 1962, a former Second World War paratrooper named William Jenkins turned up at a hospital in California complaining that the severe fits he had suffered for almost two decades were starting to get worse. Jenkins had parachuted behind enemy lines over France and engaged in hand-to-hand combat with a German soldier, who had hit him on the head with the butt of a rifle. He sought treatment after the war, but nothing seemed to work, until he arrived at the offices of a young neurosurgeon who proposed a novel and extreme procedure. Jenkins agreed to the treatment and, in the process, became an unwitting superstar of neuroscience, the primary subject of Nobel Prize–winning research, and the source of one of the most pervasive myths of modern science.

Where does originality come from and reside in our brains? I'll wager that all of us would immediately think of the "left brain/right brain" formulation that was popularized in the 1980s. For many years, there has been a widespread belief that the side of the brain that dominates our way of thinking also determines our skills, preferences, and even our personality. This is what neuroscientists call the "hemispheric model": the notion that creativity and emotion are located in the right half of the brain, while rationality and logic are situated in the left half of the brain.

Many of us have heard claims—or perhaps even made them ourselves—of being left- or right-brained, followed by a statement like "so this task really isn't my thing." Are you right-brained and trying to do your annual accounts? Think again. Or are you left-brained and trying to come up with a decent marketing design? Sorry. Doomed.

The problem is that this is almost entirely nonsense. Although the ideas have now been debunked—and indeed had never been truly supported by the science—the myth remains pervasive. But examining the roots of a lie can reveal important truths.

J. P. Guilford's work set off a revolution in psychology, behavioral science, and the then-emerging field of neuroscience, which sought to understand where convergent and divergent thinking occurred in the brain. A series of famous experiments was conducted by Roger Sperry and Michael Gazzaniga of the California Institute of Technology in the 1960s.

The problem was that while Sperry and Gazzaniga did indeed identify these patterns, the people they were researching were atypical. They were all patients with epilepsy who had undergone a radical form of surgery to divide their brains into two separate parts—so-called split-brain patients.

Cutting the connective strand between the two parts of the brain may seem like an extreme and brutal measure. That's because it was. First trialed in New York in the 1940s as a way of treating severe epilepsy, the idea was that by severing the corpus callosum, it would create a kind of neural firebreak, and prevent seizures from spreading across the brain and becoming debilitating.

But the results were not especially promising. Some patients showed some improvements, but it was not definitive enough for the practice to become widespread. It was not until 1962 that the procedure was undertaken again, on William Jenkins, in an attempt to ease his seizures. Post-surgery, Jenkins showed remarkable improvements in terms of the frequency, intensity, and duration of his seizures. But the surgery also left him with some curious effects in his thinking and behavior, which Sperry and then–graduate student Gazzaniga documented through observation and experimentation.

One of their most famous experiments related to verbal and visual reasoning. Jenkins was asked to press a button whenever he saw an image. The researchers flashed different stimuli in front of each of his eyes, which would be delivered to and processed by the opposing part of the brain (the left eye delivers information to the right side of the brain, and vice versa). Whenever stimuli were sent to his right eye—targeting the left hemisphere of his brain—Jenkins would press the button and describe clearly what he saw on the screen. But when the stimuli were sent to the left eye—targeting the right hemisphere of his brain—Jenkins would say he saw nothing. But at the same time, his left hand would press the button to indicate he had seen something. When he was given a pen in his left hand, and an image was flashed to his left eye, Jenkins would say he saw nothing. But with his left hand he would draw the object he had seen. His left and right sides literally did not know what the other was experiencing or doing.

By examining and running experiments on Jenkins and a small group of other split-brain patients over the course of the 1960s and 1970s, Sperry, Gazzaniga, and colleagues were able to show that certain cognitive functions were "lateralized," taking place in distinct halves of the brain. The left and right sides of the brains of the split-brain patients had distinctive features. For example, the left side of the brain had a stronger role in language, numerical processing, and logical reasoning; while the right was more visual and played a greater role in intuitive decision-making. They also concluded that the 1940s cohort of patients was not fully split—which explained both the lack of improvement in their seizures, and also the lack of changes to their cognitive performance after surgery.

The team's results would captivate the general public and the cre-

ative industries, and inspire a slew of books, courses, and coaching efforts focused on harnessing the so-called "creative side" of the brain.
Betty Edwards's *Drawing on the Right Side of the Brain*, published in
1979, is still the best-selling art instruction book ever written, and owes
a direct debt to the drawing experiments with Jenkins. As she puts it:

> Paradoxically, learning to draw means learning how to make a
> mental shift from L[eft]-mode to R[ight]-mode. That is what a
> person trained in drawing does, and this ability to shift thinking
> modes at will has important implications for thinking in general
> and for creative problem solving in particular.

Another bestselling creativity self-help book, Julia Cameron's
The Artist's Way, published a decade or so later, takes this further, with
vivid anthropomorphism:

> We are victims of our own internalized perfectionist, a nasty
> internal and external critic, the Censor, who resides in our left
> brain . . . Think of your Censor as a cartoon serpent, hissing
> vile things . . . *Logic* brain is our Censor . . . *Artist* brain is our
> inventor, our child . . . our creative, holistic brain . . . teach logic
> brain to stand aside and let artist brain play.

Sperry and Gazzaniga's research on split-brain patients actually
gave a much more nuanced picture, and certainly did not make overarching claims about the brains of healthy—as opposed to split-brain—
individuals. They found that the brain isn't like a single computer, with
specific sections of hardware charged with specific tasks. Instead it works
more like a network of computers connected by many cables. The connectivity between active brain regions turns out to be just as important,
if not more so, than the operation of the distinct parts. What split-brain

patient research really showed was the impact of disconnecting a large part of the network without damaging specific components.

And in these situations, there is indeed a strong degree of lateralization, with different functions being located in different sides of the brain. For those of us fortunate enough to have the network intact, lateralization is not apparent, but also completely unnecessary. This research won Sperry a share of the 1981 Nobel Prize for medicine.

Sadly, the left-brain/right-brain neuromyth remains resilient—and potentially very damaging. In one of my roles, I advise the Paris-based intergovernmental Organisation for Economic Co-operation and Development on innovation and creativity issues. A major project looked at the brain and learning, and found that many misconceptions exist among educational professionals about the brain. These so-called "neuromyths" are loosely based on scientific facts, but have adverse effects on educational practice. And left brain/right brain is one of the most prevalent of all.

Research undertaken in 2012 in the UK and the Netherlands showed that 91 percent of surveyed UK teachers and 86 percent of Dutch teachers believed that the difference between left- and right-hemisphere thinking explained individual differences among learners. Given the role of teachers in encouraging and supporting learning, this has huge implications for how students develop and learn, the effort they put in, and their levels of self-belief when being introduced to new subjects. And it is not just teachers who hold this view. In professional areas, too, people label themselves as not being creative, and in many cases relate creativity to the visual arts. This is a profound shame and is the exact opposite of what J. P. Guilford intended—and found—in his work on divergence thinking and creativity.

Fortunately, since the 2010s science has started to make something of a comeback. And Guilford's ideas are proving as pertinent to the twenty-first century as they were to the mid-twentieth. The development of techniques such as functional magnetic resonance imaging (fMRI) have enabled real-time and direct analysis of how the brain works in many different contexts and situations. Although the original applications were strictly medical, the digital revolution has led to a radical drop in the cost and improvements in the portability of brain-imaging tools and techniques. As a result, they have been applied in all kinds of contexts and situations—from traders in investment banks to Formula 1 race car drivers and surgeons in emergency situations. Thanks to these recent advances, we now understand much more than we did about the workings of our brains when performing all manner of different tasks.

Importantly, a lot of effort is being taken not to simply extrapolate from brain imaging to performance, so as to avoid a twenty-first-century retread of the left-brain/right-brain debacle. A careful effort is being made to base conclusions on robust science. The linkages between brain network performance and higher-level behaviors are being traced in a systematic fashion, thanks to an approach called connectome-based predictive modeling. This is a data-intensive approach that involves first analyzing behavioral traits of individuals independently, and then subjecting those individuals to a series of experiments where brain activity is monitored in real time while undertaking different kinds of tasks or solving different problems. The analysis then looks for patterns of behaviors and decisions that link to brain network activity.

Of particular importance for us is that there have been growing efforts to try to understand subjects' brains when undertaking original and novel thinking—specifically, while undertaking Guilford's Alternative Uses Test and its numerous variants. Researchers at Harvard University analyzed brain activity during the Alternative Uses Test. Their research showed that there are three systems within the brain that are most important for original thinking. These are the *imagination* system, the *salience* system, and the *executive* system.

The *imagination system* is integral in creative activities such as brainstorming and daydreaming. This is the system that is in control of mental simulations. It is also responsible for social cognition, such as our ability to understand what someone else may be feeling, or how to react in a particular social situation. The system is linked to exploration of and engagement with novel ideas, objects, and scenarios.

The *salience system* is in charge of monitoring internal and external stimuli: this means both our own internal stream of consciousness, as well as the external stream of information triggered by our perception of what's happening around us. It takes the massive amount of sensory information that we are constantly being bombarded with and chooses which we should pay attention to—and which we should ignore. The salience system is involved in making decisions about what information, internal or external, is most important for solving the problem at hand. It is also important for its role in switching between various neural systems, and has been shown to be vital for our capacity to have sustained focus and willpower.

Finally, the *executive system* is responsible for targeted attention, interpretation of information, and communication between the areas of the brain responsible for decision-making. This system is linked to our ability to make ideas practical and tenable, and is associated with precision, critical sensitivity, and awareness of different perspectives and audiences.

Each of these systems plays a distinctive role in our thinking processes. So, what happens in the brains of people undertaking divergent thinking tasks? One might think that it is the imagination network that matters the most. But this would be making the same kind of oversimplified interpretations as were made of the split-brain research. Instead, what the evidence shows is that it is the quality and nature of the interactions *between* these networks that matters. What the researchers found, having examined the brain functioning of subjects undergoing the Alternative Uses Test, is that none of these systems alone determines divergent thinking.

Think back to your own ideas about the paper clip. To start off with, you will most probably have come up with conventional uses, and after a certain number, more radical ideas might have emerged. This has in fact been shown by brain scans. Initially, we see coupling between the imagination and salience networks, but this is not fixed. There is then coupling between the imagination and executive networks, and then back to the imagination and salience networks. Our brains switch between generating ideas and assessing them in a dynamic fashion.

Echoing Keith Jarrett's experiences from the previous chapter, this is how these networks have been seen to interact when musicians are composing a new musical piece:

> [This] would initially require increased activity in the Imagination Network in order to evoke emotional reactivity and sensory processing necessary to produce a novel musical motif. However, once you have the beginnings of the creative lick in your head, the Salience Network would switch you over to the Executive Attention Network so you could create a working memory of the otherwise transient melody and focus on nailing the music down into a memorable cadenza.

If you've ever experienced that feeling of "firing on all cylinders" then now you have a biological explanation for it. This is because three major systems of the brain are being engaged and we feel the—literally!—heady energy and excitement. Together these three systems cover a good proportion—some 80 percent—of the brain (putting to rest, incidentally, another prevalent neuromyth, that we only use 10 percent of our brains). It turns out that among individuals with the highest divergent scores, far from being fixed in one hemisphere, the whole brain is actively employed and in a variety of different ways.

In many situations—and for many people—these systems often work in opposition to one another. It also turns out that many of us, through habit or laziness, have learned to rely on certain network interactions more than others when making decisions. Over time, certain systems can become strong—even entrenched—preferences.

When undertaking divergent thinking tasks, regions within the three systems are networked together. Original thought relies on the constant switching of connections between the brainstorming capabilities of the imagination system, the internal-external monitoring and learning of the salience system, and the decision-making and detail-oriented executive system. The connections between these three networks, and the speed at which they interact, are in fact the best and most robust predictors of how good someone will be at divergent thinking. Individuals with more connections between these networks are more likely to think outside the box, and to think of new solutions and possibilities. They are also more likely to be able to be systematic in how the different networks are harnessed, and in what order.

Among the most effective divergent thinkers, there is a clear ability to simultaneously engage all three networks, and to do so more efficiently than the rest of us. The participants who produced the greatest number of original responses in the test also had greater network efficiency: Their brains demonstrated the capacity to traverse the networks with a smaller number of steps than less-original thinkers.

The connectome-based predictive modeling approach is starting to show that these neural connections can form the basis of variations in decision-making skills and style. As we will see in part II, Upshifters' skills correspond closely with these systems.

As a nice coda to the left-brain/right-brain myth, when the Harvard research on the networked neuroscience of divergence was accepted for publication in the prestigious *Proceedings of the National Academy of Sciences* in 2018, the editorial board member who accepted it was none other than Michael Gazzaniga, who undertook the first experiments on William Jenkins alongside Roger Sperry.

Just as our preferences and habits are not fixed, neither are the neural networks that underpin them. J. P. Guilford was always adamant that creative capabilities could be developed over time. It turns out that this is true, not just at a behavioral level but at the level of neural architecture, and Guilford's ideas and techniques again have played a central role in these findings.

Longitudinal research undertaken in China has examined how brain functions and structures underlying original thinking might change over time. Participants were selected and tested using the Alternative Uses Test, with the researchers ensuring that at the start of the experiment all participants had very similar divergent thinking capabilities and neural patterns.

The test participants were then divided into two groups, with one then involved in twenty sessions of divergent thinking training over the course of a month. In the thirty-minute sessions, each person was invited to undertake the Alternative Uses Test as well as a number of other cognitive-training simulations (improving a specific set of products, exploring the implications of different social situations). Their brain activity was analyzed during the task and after. The other half of the participants formed a control group that was analyzed at the end of the period.

Over the course of just a month, using only these simple divergent thinking techniques, the test groups showed a marked improvement in both the fluency and originality of their thinking. Compared to the control group, they came up with more ideas and the ideas were more original. But, more remarkable than this, the test group participants also showed significant structural and functional changes at a neurological level. That training can contribute to neuroplasticity is not new—trained taxi drivers famously have enlarged spatial recognition regions—but this is one of the few that have showed not only that creativity can be taught but also the results can be seen in our gray matter. The changes in the test group could be observed in each of the three neural systems I described above: in terms of how they functioned, as well as, importantly, how well they integrated with one another.

In the last chapter, we learned how our perceptions of stress trigger our biological responses. Here is the equivalent in original thinking: our way of thinking can change the biology of our brains. In this case, the changes are more fundamental than simply altering the chemistry of the brain; they have a direct influence on neural structure and function. The researchers also found that there was a connection between participants' stress appraisals and how well their brains worked as a whole network. Thinking back to the chart in the introduction, those participants who were in an optimal state of "arousal"—in the challenge state, as opposed to being underloaded or overloaded—were also more likely to have their whole brain engaged, and more likely to come up with a greater number of original ideas.

So why are these findings not more established and utilized across our society? Why are neuromyths like left brain/right brain so persistent and pervasive, while the genuine, more empowering scientific conclusions are confined to academic circles?

My personal and professional experiences in humanitarian work suggest that the biggest problem with divergent thinking is that not only do most institutions tend not to reward it but also many actively punish it. Ironically one of the worst culprits of all is the modern educational system.

LEARNING WHAT NO ONE KNOWS HOW TO DO

My son, Koby, always had a somewhat ambiguous relationship with primary school. He is thoughtful and hardworking, and is well liked by his teachers. He loves learning and enjoys being with his peers. But there is also an underlying frustration he voiced from time to time. Once he said to me, "School only teaches you to do the things people already know how to do. Where do you learn how to do things no one knows how to do?"

It's a question that would no doubt make Upshifters like Chesley Sullenberger smile. It has also obsessed educators around the world, including a noted creativity speaker and advocate, the late, great Sir Ken Robinson. In a world-famous TED Talk, "Do Schools Kill Creativity?," Robinson highlighted divergent thinking as an essential capacity for creativity. Referencing Guilford's Alternative Uses Test—and the paper clip—he described how divergent thinking levels changed over time among a cohort of children from kindergarten to high schoolers:

> At the age of five years, 98 percent of children were "genius level" in terms of the number of different uses they could generate.
>
> Five years later, when they were age ten, those at genius level had dropped to 30 percent.
>
> After another five years, the number of divergent thinking geniuses had fallen further still, to 12 percent.
>
> The same test, administered to a much wider but unrelated cohort of 280,000 adults, found that only 2 percent registered at the genius level for divergent thinking.

If you are anything like me, you will be gasping, "Why?"

Robinson convincingly argued that the main reason for this jaw-dropping decline is that children have their creativity educated out of them. He identified a conveyor-belt-style school system built on the principle that problems have "one answer, at the back of the book—but don't look."

As Koby would no doubt attest, teaching has tended to skew toward teaching routine tasks that follow a step-by-step process, rather than encouraging complex and creative problem-solving. As a result, we "surrender our creativity" through—and to—the educational process.

Some of us, teachers especially, may be feeling somewhat affronted by this. Of course children are encouraged to think openly and creatively in classrooms, and there is no doubt that teachers naturally draw upon a range of approaches and techniques. The problem is less that there is no exposure to divergent thinking. It is more that, as George Land—who led the creativity study cited by Robinson—argued, the approaches are "mashed up together." So, echoing the left-brain/right-brain neuromyth, we are taught to use convergent thinking in math, grammar, and languages and divergent thinking in aspects of art, design, and theater. As a result, our brains are literally all over the place, and we cannot use our divergent capabilities to their full potential: "competing neurons in the brain will be fighting each other, and it is as if your mind is having a shouting match with itself."

But again this should not lead us to be disheartened. As we saw earlier, we can learn divergence—and we can also *re*learn it. Numerous studies have shown that both children and adults can develop their divergent thinking skills through a range of techniques.

In schools, the methods can be very simple and powerful. In one study, teachers were asked to pose divergent thinking challenges on a daily basis to elementary school kids. These were open-ended questions—that don't have a right or wrong answer—that deliberately asked the children to think in a more open way about what might happen in a variety of scenarios. They also put them through the Alternative Uses Test before and after the two-month-long experiment, and identified examples of divergent thinking from all the classes over the same time period.

The classes that used the divergent thinking tools came up with ten times more creative ideas over the two-month period. They also did better than the control classes on every measure of the Alternative Uses Test (flexibility, originality, and so on). The conclusion was that "young children can realize dramatic increases [in innovative thinking] when repeatedly exposed to divergent thinking situations."

For adults, we have already seen that cognitive simulations can help to develop our original thinking capacities, with just a month of training in the cross-China study leading to changes both at a behavioral and a neural level. Other simulations are less laboratory and more real-world.

Both teenagers and adults participating in improvisational theater classes showed measurably improved levels of divergent thinking over time: "Improv promotes peoples' ability to find creative and resourceful solutions and that therefore it can help us to think in more diverse ways or even break away from ingrained patterns of behavior."

All this may sound very good. But how does it actually help us in "moment of truth" situations like those faced by Upshifters? In fact, a large proportion of the divergent thinking experiments that have been undertaken focus on putting people under different kinds of constraints to assess their performance under pressure. For example, firefighters and other emergency-rescue specialists in Australia were asked to think about novel approaches to disaster response with a blank sheet of paper. They were then given a series of real-life constraints, in terms of resources and contexts, and showed a tenfold increase in their creative output.

In a very different professional setting, staff in a cereal-manufacturing organization were assessed for their creative performance when subjected to different levels of time pressure. Looking at almost two hundred staff members, it turned out that their performance, assessed in terms of the quantity and quality of their original ideas, increased with pressure—up to a point.

As with the stress/performance trade-off we learned about in the previous chapter, an inverted U shape was identified. Employees exhibited relatively high creativity when they experienced intermediate creative time pressure, as long as two additional factors were taken into account.

First, the individuals needed to be open to new experiences under stress—"a constant quest of unfamiliar situations characterized by a high degree of novelty." Participants who saw stress as a potential catalyst for performance were, when induced, more likely to come up with original solutions, and a greater number of them. Think back to Sully and his colleagues striving to make each flight "better than the previous one."

Second, the inverted U was especially apparent where there was social support for original thinking from peers and managers. It is not just individual characteristics that shape whether or not you can be original under pressure: it is your social environment. Specifically, it is about how those around you see originality as an "expected and valued aspect of their performance." Remember Jarrett feeling buoyed by the Cologne audience's expectations. Remember teachers who have inspired you to learn and explore. By supporting our space for originality, those around us increase the likelihood of click moments.

Not only can social support allow us to be more original when facing pressure, but it can also help us deal with pressure more effectively. Research on medical residents at Johns Hopkins Hospital in Baltimore showed that teams' learning behaviors—how medical colleagues gather information, reflect on experience, share knowledge, and generate new ideas—not only shape the performance of the team in high-pressure situations, but also lead to significantly lower levels of burnout among team members.

Thinking about all of this and, going back to my son's frustrations, what might a school that placed a genuine importance on originality look like? Ken Robinson wrote a treatise on creative schools, based on his appraisal of the challenges faced across many developed countries. He noted that politicians had retreated into "exam-factory education"—what one observer described as a "self-defeating cul-de-sac of testing and assessment"—they had become breeding grounds for convergent thinking.

This was exactly the problem with the testing of fighter pilots J. P. Guilford identified half a century earlier; schools weren't teaching kids to survive and thrive—they were teaching them to crash and burn.

Echoing Guilford, Robinson advocated for schools that personalized learning, appreciated the diversity of intelligence, adapted teaching schedules to different learning speeds, allowed learners to pursue their interests and strengths, and used different models of assessment based more on trust and empowerment. In other words, schools needed to spend more time improving abilities rather than proving them. As Robinson argued, creative schools would have to start "teaching children how to employ their natural creativity and entrepreneurial talents—the precise talents that might insulate them against the unpredictability of the future in all parts of the world."

In the final part of our journey on original thinking, I want to turn not to the future, but to the deep past of our species, and show that these "precise talents" proved vital for our very survival and existence.

PAINTING THE CAVE RED

In known human history, the cape of South Africa has shaped our imagination. It has represented the end of many explorers' dreams, a mirage, something untraversable, and, eventually, part of our voyage to modernity.

But it also played an essential role in human prehistory. Looking at the world today, teeming with people, with millions of species on the edge of human-driven extinction, it might seem strange that we were once an endangered species ourselves. And yet the genetic fingerprints each of us carries within tell a story about where we came from. We span the Earth in huge numbers and yet our genetic diversity is tiny: much lower than species with smaller numbers and narrower geographic ranges. Remarkably, the entirety of the human species has less genetic diversity than tribes of chimpanzees living on opposite sides of a single river in central Africa.

Archaeologists have linked this genetic bottleneck to an ice age that lasted from around 200,000 to 125,000 years ago. This climate catastrophe is inscribed in our genes. During this period, the number of humans dropped dramatically, from over ten thousand to just a few hundred. Given the narrow genetic diversity of modern humans, it seems likely that the main progenitor population was a single group that survived in one part of Africa, then spread outward, mixing with other populations as it moved.

Which brings us to the Cape Floral area, one of the most southerly points of the African continent. It is both beautiful and rugged. Rocks and vegetation seem to be in a constant tussle, with one winning here, and the other there. The cliffs overlooking the churning Indian Ocean are rich in plants. Indeed, this thin strip of land contains the highest diversity of flora for its size in the world. It constitutes less than 1 percent of the area of the African continent but is home to an astonishing 20 percent of all of its plant species. The cliffs are also pockmarked with large caves, including one called Blombos Cave. In this modest shelter, archaeologists have made two sets of discoveries that are at the heart of why this tip of South Africa has become so important to understanding the story of our species. It is increasingly viewed as the most likely home of our early common ancestors. And it is also seen as a new earthly Eden: the literal birthplace of original thinking.

Anatomically modern humans lived in Africa as early as two hundred thousand years ago. But the question of when we developed modern mental capabilities was until recently the subject of intense debate. Findings regarding the early humans living in Blombos Cave and others like it on Cape Floral have helped to resolve this debate—by upending it.

Many of us are aware of early cave paintings, most of which have been found in Europe, and date back some forty thousand years. In fact, there has been a whole host of discoveries dated to the same period, all in Europe, which archaeologists have long taken as markers for the earliest cognitively modern human behavior. They include standardization of artifacts, blade technologies, worked bone, personal ornaments, structured living spaces, and the emergence of art and symbolic images.

The prehistoric inhabitants of Blombos Cave have stuck two fingers up to this archaeologically Eurocentric viewpoint. They crafted composite weapons from stone, wood, and shell from seventy thousand years ago—thirty millennia earlier than previously believed. A hundred thousand years ago, the cave's residents were already making tools that clearly demonstrate the use of fire to heat and flake their edges to make them sharper, sixty millennia before the evidence in the European record. Traces of burned vegetation suggest that these ancient hunter-gatherers figured out that by clearing land they could encourage quicker growth of edible roots and tubers eighty thousand years before the agricultural revolution took hold. And there is strong evidence of shellfish foraging and fishing.

The people of Blombos Cave and the surrounding areas provide us with the clearest evidence of early human symbolic thought and learning. In 2018, a dig led by archaeologist Christopher Henshilwood at Blombos Cave revealed the earliest known human drawing, a red ocher etching on a rock, which predates European cave art by at least forty thousand years. Perforated ornamental ostrich shells have since been identified as some of the earliest human jewelry. The inhabitants even painted the walls of their caves red. The most recent research on these cultural artifacts and how they evolved over time shows that the symbols used—whether engravings on shells, carvings on rocks, or paintings—became more sophisticated over time: "more salient, memorable, reproducible, and expressive of style and human intent."

Their use of tools shows that they had an understanding of the properties of materials, and how these could be combined. Their workmanship demonstrated an ability to plan. Their exploitation of natural resources for food is even more impressive. Fishing and coastal foraging would have required an intelligent understanding of tidal patterns, and the use of lunar cycles to plan and coordinate life-threatening expeditions out onto the treacherous coastal waters.

Their use of art and decorations, and red ocher for painting the walls of their cave, demonstrate their symbolic reasoning. The identity of this group mattered to them. *What they did* and *who they were* became intertwined. And they represented who they were in all kinds of novel ways. They forged not just a way of living, but a sense of collective meaning. Their advanced intellect and collective identity contributed significantly to the survival of the species.

These findings about the inhabitants of Blombos Caves "push back by at least 20,000 or 30,000 years" the evidence for when our species evolved complex cognition. We have effectively extended our understanding of the development of the human mind.

The questions then are: how did this happen, and why? Evidence suggests that dietary practices were an important catalyst. The diversity of plants here on Cape Floral provides one clue. There are more *geophytes*—plants that store energy in underground structures—than anywhere else on the planet. The Blombos inhabitants happened to live in the world's largest larder of root vegetables. Energy rich, and able to survive harsh environmental conditions, geophytes are also low in fiber, making them easier for children to digest. Seafood also served as a nutritional trigger, providing vital fatty acids that are needed to fuel thinking. As one of Henshilwood's colleagues puts it: "[Nutrition] is the evolutionary driving force . . . It is sucking people into being more cognitively aware, faster-wired, faster-brained, smarter."

These dietary practices and neural developments were catalyzed by external pressures. The environmental changes that drove the population bottleneck also created opportunities for adaptation and change. By studying the populations on the coastal regions of southern Africa, researchers have found that there was a pattern of culturally driven adaptation that enabled the Blombos people and others like them to exploit new and changing ecological niches. The complex technologies and symbolic practices of the Blombos populations have been directly linked to novel adaptations made in response to the environmental conditions they faced.

What is especially interesting is how the archaeological record shows how changes in their practices corresponded to changes in environmental pressures they faced. Specifically, a number of complex techniques were replaced during periods of pronounced aridification with more flexible adaptations that enabled the groups to exploit broader ecological niches. When new niches appeared, the Cape humans did not come up with entirely new approaches. Instead they used what the archaeologists have described as the earliest examples of "cultural exaptation," defined as the "use of existing skills, techniques, and ideas in new ways." This is the prehistoric equivalent of the Alternative Uses Test: their response to environmental pressures was divergent thinking. The Blombos were the original Upshifters.

Unlike much prehistoric archaeology, which is hard to move beyond conjecture and coincidence, there is a more solid evidence base reinforcing these findings. When archaeologists sought to understand exactly how these practices influenced the inhabitants of Blombos, they used the existing physical evidence to try to extrapolate neuroscientific lessons. In one remarkable set of experiments, they subjected modern-day research subjects to the visual and symbolic products—paintings, tools, and processes—of the Blombos humans while measuring brain activity using fMRI scanners. The Blombos artifacts and processes led modern humans to engage the same parts of the brain that are engaged in divergent thinking tests.

Researchers have concluded that Blombos gives the earliest evidence for "enhanced executive functions" of the brain that "encompass innovative thought and imply the ability for complex goal-directed actions, flexibility in problem-solving, task switching, response inhibition, and planning over long distances or time." Christopher Henshilwood argues that all of this suggests these early ancestors were not merely survivors—but that they lived through a "dynamic period of innovation."

In my view, Blombos should be seen as important a site for human innovation as early Mesopotamia, ninth-century Baghdad, fifteenth-century Florence, nineteenth-century England, or twentieth-century California. It was the site of a *cognitive* revolution, which arguably had a greater impact on the future of our species than the agricultural, industrial, or digital revolutions that would follow.

But if this is indeed the earliest evidence for cognitively modern humans—as demonstrated through symbolic behavior, complex technologies, and divergent thinking—then why do we tend to see east Africa as the cradle of humanity? The latest cross-continental genetic analysis of human remains suggests that while the "glacial refuge" of Blombos was indeed the site of the flowering of these distinctive features, the archaeological records suggest that soon after seventy thousand years ago, there was a movement from southern to east Africa, and this immediately preceded the first major human demographic expansion, which led to our species peopling the rest of the world. Moreover, the evidence also suggests that the cultural innovations from the southern African refuge were in fact the trigger for these expansions. East

Africa may be the cradle of modern humanity, but southern Africa was our moment of conception.

The Blombos peoples' creative potential was passed down to every human being alive today. We all have the ability to use pressure and stress as a catalyst for creative transformation. Pressure and even calamity can be used as drivers of ingenuity. It helped us survive the ice ages and turned us into modern humans. And many tens of thousands of years later, Upshifters around the world use the same skills to navigate pressure and stress in divergent ways, not just when facing epic and extraordinary stresses and pressures, but also in everyday settings.

To give the final word in this chapter to J. P. Guilford himself: "To live is to have problems—and to solve problems is to grow creatively."

3

THE STRENGTH OF PURPOSE

WHAT CAN NEVER BE TAKEN AWAY

SHINING THROUGH THE TROUBLES

July 21, 1972, was a pleasantly warm afternoon in Belfast, Northern Ireland. At the hairdressing salon over the road from the Royal Victoria Hospital, the chief casualty nurse, Sister Kate O'Hanlon, and one of her emergency department colleagues were midcurl.

At 2:10 p.m. the first bomb exploded, at the Smithfield bus station. It was followed in quick succession by nineteen more across the city with less than ninety minutes from first to last. As the devastation from the blasts accumulated, the city center became filled with smoke and screams. The targets seemed random: parked cars, taxi firm headquarters, train station platforms, bus depots, hotel lobbies, shopping districts, banks, bridges, gas stations.

As soon as she heard the first blast, Sister O'Hanlon and her colleague, their hair wet and still in rollers, ran to the hospital. Within minutes the first bodies and casualties were being wheeled in. Despite years of frontline experience, Sister O'Hanlon was unprepared for the level of injuries and trauma. The carnage on the streets was so horrific that human remains and body parts were collected and brought to the hospital in plastic bags, which were then piled up on trollies in corridors. This was Bloody Friday.

That year would see the worst of the sectarian conflict in Northern Ireland known as the Troubles: five hundred people killed and over twenty thousand injured. The majority of casualties occurred within a mile of the Royal Victoria, the largest hospital in Belfast, on the west side of the city. On that grim day, 130 people were seriously injured, and nine killed. More than half of the casualties from the bombings ended up under Sister O'Hanlon's care.

Bloody Friday would see an abrupt turnaround in what had been growing public sympathy for the Provisional Irish Republican Army (IRA), which was widely believed to have carried out the attacks by explicitly targeting innocent civilians. The city was still mourning the brutal Bloody Sunday killings by British paratroopers six months earlier, and trust in the British Army was running at an all-time low. After the IRA attacks of Bloody Friday, it seemed that no one was really standing with the people of Belfast.

Almost no one.

After the attacks, the Royal Victoria was described in a popular Belfast broadsheet as "no ordinary hospital . . . right in the middle of the battle." It was one of the few institutions that was trusted by all parties to the unfolding intractable conflict. We get a clue as to why this was from Sister O'Hanlon's autobiography:

> One day somebody came in shot and we [also] had the man who shot him in . . . It doesn't matter because once they come through the door they are a patient . . . All we want to know is what really happened in relation to the injuries. We are not interested in the politics.

The determination of the Royal Victoria staff to fulfill this humanitarian purpose—to maintain humanity, values, and independence in the face of a bitter conflict, to treat everyone the same—was unwavering.

But the Royal Victoria was more than its humanitarian accomplishments. The hospital became a hub for innovation in trauma medicine: triage methods, tools for preventing lung collapse in bomb blast victims, new kinds of splints, even fusing dental techniques with critical cranial surgery. These approaches are even now used in emergency rooms and war zones around the world, including in Afghanistan, Iraq, and Syria.

Not only did this remarkable work happen in the midst of conflict, it is apparent that the flourishing of creativity happened *because* of the conflict. According to a BBC report, many of the Royal Victoria practitioners "became world-famous for the techniques and procedures that they *invented under the intense pressure* of dealing with events like Bloody Friday" [emphasis added].

But in 1969, just three years earlier, very few of the Royal Victoria staff had even seen a gunshot wound. One nurse explained that, prior to the start of the Troubles, the emergency rooms were seen as an easy shift, often staffed only by students.

At that time, the UK's National Health Service (NHS) was not prepared for dealing with serious long-term violent conflict. Large-scale emergency preparation focused on the kinds of treatments that were required after industrial and transportation accidents, or natural hazards such as flooding.

Looking at the other hospitals in Belfast during the same period, there was nothing like the flowering of innovation and creativity seen at the Royal Victoria. Despite the trauma, devastation, and political complexity, Sister O'Hanlon and her colleagues found a way to "shine during the worst of the Troubles." So, what made the difference at the Royal Victoria? What was the source of the light?

CRISES ARE THE GREATEST CLASSROOMS

We have already seen how we can deal with stress and pressure with the mentality of facing a challenge as opposed to a threat, and how we can turn to *originality* when faced with novel and unfamiliar problems. The third and final ingredient of Upshifting is a sense of *purpose*; something that marks out the extraordinary achievements of individuals and teams like those at the Royal Victoria Belfast when experiencing stress and pressure.

Harvard historian Nancy Koehn has undertaken research on more than a hundred leaders from around the world, men and women—political leaders like Abraham Lincoln, moral leaders like Rachel Carson, operational leaders like explorer Ernest Shackleton, business leaders like Howard Schultz of Starbucks. All of the leaders she writes about faced a "boundary situation"—which is akin to the "moment of truth" Captain Sullenberger experienced back in chapter 2. They faced these situations by "taking a single step forward into the turbulence and then the next step after that."

In an interview, she deepens her thinking on this:

> This is the realization of many of us who have been through a major challenge, that there is something fertile in all this adversity, all this confusion, all this ambiguity, and at times borderline despair, in that "I can learn something important for myself of what I might be."

What was that fertile thing to be found in adversity and pressure? Koehn's in-depth investigation of leaders who were "forged during crises" has identified a *sense of purpose* as being the thing that really mattered. In essence, this means three things:

1. Identifying and articulating goals, values, and principles that guide actions, which helps to strengthen resolve and persistence in

pursuit of those goals, and helps to overcome fears and anxieties about the seeming impossibility of those goals.

2. Reframing and redefining the problems that are faced in the context of the overarching sense of purpose and taking a consciously experimental approach to developing possible solutions.

3. Bringing people together as a cohesive whole around that purpose based on shared passion, mutual trust, and solidarity.

While the phrase "a sense of purpose" can evoke pop psychology or managerialist self-help jargon—and I must admit to having had this response myself—the reality could not be further from the truth. There is ever-growing evidence that a sense of purpose can have a profound effect on our lives that goes well beyond what we might expect. It is also one of the three essential elements of Upshifting.

The scientific interest in understanding purpose began in earnest with the work of Viktor Frankl, a psychiatrist and Holocaust survivor. Frankl not only lived through the horrors of Auschwitz himself—where his pregnant wife, father, mother, and brother were all murdered—but also took on the role of therapist for his fellow condemned prisoners.

Consider the enormity of the situation in which he found himself, and having the emotional strength and wherewithal and mental presence to even attempt to provide psychological support to others around him. In his subsequent reflections, Frankl came to the following conclusion about what our needs are in such contexts—and by extension, in our lives as a whole: "What man actually needs is not a tensionless state but rather the striving and struggling for some goal worthy of him. What he needs is not the discharge of tension at any cost, but *the call of a potential meaning waiting to be fulfilled* by him" [emphasis added].

This is a perfect description, by the way, of the inverted U of the curve—where we don't want either boredom (a tensionless state) or its opposite (discharge of tension at any cost) but the Upshift zone in the middle: striving and struggling. And there is no better definition to my mind of a sense of purpose than "the call of a potential meaning waiting to be fulfilled."

That purpose was found to be significant even in the most dehumanizing conditions imaginable was a message that resonated far and wide. Frankl's book *Man's Search for Meaning* was listed in the 1990s by the Library of Congress as one of the ten most influential books in the US.

A huge amount of empirical evidence has built up over the years that supports Frankl's arguments. Medical research has proven that a greater sense of purpose can help us live longer, happier, healthier lives. Life purpose is linked to lower incidence of acute and chronic diseases. Data from the previously mentioned US-based Midlife survey, which started in 1995, found that over a ten-year period, those people with a greater sense of purpose—having meaning and direction in their lives—had far lower levels of physiological stress-related impacts on their bodies. This was proved with reference to an astonishing number of indicators including cardiac health, BMI, metabolism, cholesterol, blood sugar, and cortisol levels. A sense of purpose has also been linked to greater cognitive performance, well-being, and happiness. A separate study of seven thousand participants found that a sense of life purpose was significantly associated with lower levels of mortality among sixty-year-olds. A sense of purpose can also be seen as a vital resource to draw upon for the other Upshift ingredients: numerous studies have shown that it underpins our ability to establish and maintain a challenge mentality in the face of pressure, and also to pursue original and creative approaches.

The evidence that a sense of purpose is essential for dealing with pressure and stress goes beyond the individual to groups and organizations. One of the most influential scholars on creativity in the world is Teresa Amabile at Harvard University. Over four decades, she has shaped how we think about creativity, and its enablers and constraints.

Defining creativity as the "production of ideas or solutions that are both novel and useful," Amabile has done extensive work on the conditions in which creativity flourishes—or does not. She has found four distinct ways in which creativity emerges from the interaction between social environment and pressure:

1. *Autopilot:* The group works under low pressure, and meaning/purpose is low.
2. *Treadmill:* The group works under high pressure, and meaning/purpose is low.
3. *Expedition:* The group works under low pressure, and meaning/purpose is high.
4. *Mission:* The group works under high pressure, and meaning/purpose is high.

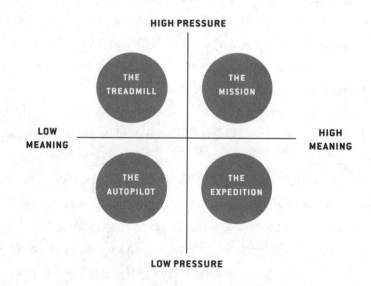

As Amabile sees it, the *ideal* scenario for creativity to flourish is in expedition-style jobs. But as we saw in the last chapter, we cannot choose when we need to be creative. Amabile found that in situations where pressure can't be avoided, you can get people to be creative by instilling a meaningful purpose, where they have a sense that there is an important, urgent need for the work they do. As Amabile puts it, "Often we have no choice about being under the gun. But if we want to be creative, we have to learn to dodge the bullets."

Much like the casualty team at the Royal Victoria during the Troubles. Their remarkable achievements beg the question: how did a group of individuals facing a complex, unpredictable set of challenges manage to develop a whole series of creative and innovative responses that would not only save lives and preserve humanity, but also go on to influence practices around the world? It turns out the sense of purpose of the Royal Victoria team was a big part of what made them all different.

Think back to Sister O'Hanlon determinedly treating all patients the same, regardless of politics or whether they were perpetrator or victim. She would always brief her staff to do the very best that they could do, and to come up with new ways of doing things when the old ones did not work. Her constant refrain was, "Imagine it is your family member being wheeled in—wouldn't you want them to have the best possible treatment in the world?" This purpose-driven creativity was infectious, and the Royal Victoria was blessed to have other people with the same focus.

This says a great deal about the importance of how closely the team held their sense of shared humanity in spite of the horror, and it exemplifies the way all of the staff at the Royal Victoria thought and worked.

At the time of the Troubles, there was an ongoing debate about the staffing of such emergency departments in the NHS and further afield. There were very few in the UK and even fewer in Europe that were run by senior doctors who also worked in them. One of these rare exceptions was William Rutherford, the senior doctor in charge of the emergency department at the Royal Victoria. He pressed for the recognition of emergency medicine as a bona fide specialty, and when the first emergency consultant posts were trialed across the NHS, he was one of the very first to hold such a position. Later on, he would coauthor the first major UK textbook on emergency medicine. Rutherford's own mentality was critical: that of "a man who constantly looked for ways to improve the care he could give patients."

During his time leading the department—which included the worst of the Troubles—Rutherford authored or coauthored no less than forty-seven academic and medical papers. To have this dedication to science and learning despite the ongoing crisis is nothing short of remarkable: "He wrote about bomb blasts, gunshot injuries, and the damage done by rubber bullets, injuries which were cared for in his department in the Troubles." The plans for managing disasters Rutherford developed were commended for saving many lives on Bloody Friday and became widely adopted across the UK. And nursing advances held the whole system together. The team led by the indomitable Sister O'Hanlon was acclaimed for its ability to provide support and manage triage in even the most complex of crises.

Even more than the technical excellence of the people, though, was the collective mentality of the staff. Kate O'Hanlon's autobiography dedicates an entire chapter to "Coping with Stress." She notes: "There was no counselling, but we coped because we were all together. You had porters and domestics and nurses and doctors and consultants, and we would all talk together. There was a great team spirit."

This collective purpose wasn't just good for morale; it also greatly enhanced their creative processes. In one of the regular cross-disciplinary, hierarchy-spanning meetings that were held, the senior neurosurgeon at the Royal Victoria had a chance discussion with one of the orthodontists. Their conversation developed and eventually saw the development of titanium cranioplasty, a method adopted from dentistry to use fine metal to quickly repair the skulls of patients that had been critically damaged by bullet and projectile trauma. This was such a noteworthy achievement in the treatment of brain injuries that one of their prototypes forms is part of the permanent medical exhibition at the London Science Museum.

Just like the more famous leaders Nancy Koehn describes as being forged in crises, the leaders at the Royal Victoria—Kate O'Hanlon, William Rutherford, and others—instilled a sense of purpose in those around them. This played a vital role in the radical transformation of the Royal Victoria's emergency department from a quiet unit dealing with everyday accidents to a world-leading conflict-trauma department in just a few years.

They did so not by specifying or directing the change, but by creating the conditions that allowed for the emergence of such change, and exemplifying that selfsame change in their own actions.

Rutherford, remarked that running the department was very simple: "You have to love everybody, you have to listen to everybody, and if in doubt, you do what Sister O'Hanlon tells you."

These efforts were in keeping with the work of Meg Wheatley, a US leadership specialist who argues that in "troubled, uncertain times, we don't need more command and control; we need better means to engage everyone's intelligence in solving challenges and crises as they arise."

By her own account, there were only two times that Sister O'Hanlon resorted to alternative means of looking after herself and her team. Once, in 1971, after the bombing of the well-known Belfast pub McGurk's, at the recommendation of the hospital psychologist she shared a Valium tablet with another nurse. But it had no effect on their state of mind, and they never tried it again.

Another time, the Royal Victoria was invaded by a group of women who mistakenly thought that their husbands—prisoners at the nearby Long Kesh Detention Centre—had rioted, and that the seriously injured cases had been brought to the Royal Victoria. Sister O'Hanlon and her colleagues locked themselves in a broom cupboard and passed around a bottle of whiskey until the storm had passed.

Koehn, in her analyses of leaders forged in crises, talks about the fact that many had "small props, behaviors and tactics to keep on keeping on": Abraham Lincoln had filthy jokes and songs; the radical environmentalist Rachel Carson relied on her cats as a source of comfort. At the Royal Victoria, they had one another and their shared purpose. And the occasional glug of whiskey.

The sad reality is that most strategies and missions under pressure never come close to this. Many companies and organizations don't even try. And even when they do, many managers often unwittingly undermine the work of their team members. They do this in a number of predictable ways: by dismissing the importance of more junior teams' work; by destroying any sense of ownership by switching people around regularly, so work is never finished; by changing goals constantly, so that work is felt to be wasted; and by not keeping teams in the loop with changing priorities and thinking. Bad management can be hard at the best of times. But in the face of pressure and crisis, it can be toxic.

For aid workers, bad managers can actually cause more stress than the disasters and wars to which they are responding. For the most part, these are people who subscribe to the "great person hypothesis." They manage by command and control, seeking to direct and give orders, and believe they are somehow chosen to make a difference in the world. In contrast, good managers, who can make the difference between success and failure, are remembered not for what they did themselves, but for what they made others believe, and how they provided the space for others to excel and to transcend crises. Indeed, research on disaster responders has shown that with the right kind of leadership, the intensity of crises can be used as the basis for post-traumatic growth.

I want to turn now to show the role that purpose plays for a group of people who have been quietly using it to transform the world since time immemorial: people living with disabilities.

THE ORIGINAL LIFEHACKERS

Question: What do treatments for polio, touch screens, electric tooth-brushes, and typewriters have in common?

Answer: They all came about thanks to the efforts of people with disabilities to overcome their constraints.

One of the most iconic Upshifters in history is the US president Franklin D. Roosevelt. With his iconic phrase "We have nothing to fear but fear itself," he arguably brought Upshifting to an entire nation. But this was not simply because of his innate leadership and influence. Struck by polio in the early 1920s, aged thirty-nine, he was paralyzed from the waist down. After years of therapy, Roosevelt taught himself to walk short distances while wearing iron braces on his hips and legs, with a cane on one side and supported by a sturdy person on the other. This led Roosevelt to a lifetime of campaigning and fundraising for better preventative measures and treatments for polio, setting up what would become the National Foundation for Infantile Paralysis.

It was originally funded by wealthy benefactors, but the incoming monies could not stretch to meet the considerable demands. During the Great Depression of the 1930s, the toll of polio on American children skyrocketed, and Roosevelt appealed to the general public, with a call for a "March of Dimes":

> Those who today are fortunate in being in full possession of their muscular power naturally do not understand what it means to a human being paralyzed by this disease to have that powerlessness lifted even to a small degree. It means the difference between a human being dependent on others, and an individual who can be

wholly independent. The public has little conception of the patience and time and expense necessary to accomplish such results. But the results are of the utmost importance to the individual.

The public took his deeply personal appeal seriously, flooding the White House with a silver wave of almost three million dimes. Not only did this set the model for many other health fundraising campaigns in subsequent years, but the polio foundation would directly fund both the iron lung technology in the 1940s and the first successful polio vaccine in the 1950s.

People with disabilities have long played a central role in transformational innovations, though they are frequently—and mistakenly—viewed as "inspiration" for these processes rather than active participants in them. The reality, as disability rights advocate Liz Jackson argues, is that people with disabilities "were the original *lifehackers*. We spend our lives cultivating an intuitive creativity because we're forced to navigate a world that's not built for our bodies."

A lifehack is a clever yet nonobvious way to solve a problem or do something, and lifehackers are people who "operate skillfully and inventively, moderating and adapting tips and schemes." Life for many people with disabilities needs to be hacked by necessity: "[they] are often outstanding problem solvers because they have to be." Extensive research points to myriad individualized and ingenious ways people address physical, social, and political aspects of living with a physical impairment on their own terms.

Central to this is a sense of purpose. At the heart of many disabled innovators' efforts there is an unwavering belief that even though their disabilities may be seen by some as negative and traumatic, they can also be a source of positive change. Roosevelt was no exception. At the entrance to his memorial in Washington, DC, there is a statue of him in a wheelchair, paid for thanks to the fundraising efforts of the

US National Organization on Disability. On it is a quote from his wife, Eleanor, which states: "Franklin's illness . . . gave him the strength and the courage he had not had before. He had to think out the fundamentals of living and learn the greatest of all lessons: *infinite patience and never-ending persistence*" (emphasis added).

This "greatest of lessons" is something that comes up time and again in research on how people with disabilities have overcome their impairments. Research in Italy on young paraplegics—generally people who had lost the use of their limbs due to accidents—showed that despite the sense of loss that the accidents generated, there were a large proportion of people who saw the event as presenting them with very clear goals and a clarity of purpose that came from the process of learning to master their new limits and impairments. As one of the young men interviewed put it:

> When I became paraplegic, it was like being born again. I had to learn from scratch everything I used to know, but in a different way . . . It took commitment, willpower, and patience . . . As far as the future is concerned, I hope to keep improving, to keep breaking through the limitations of my handicap . . . Everyone must have a purpose . . . These improvements have become my life goal.

In the disability lifehacker movement, such improvements see people spending time rethinking existing tools and approaches to enhance their usability, accessibility, and usefulness. The ingenuity of people with disabilities to solve the problems they face, by finding new ways to meet individual needs, wants, and capabilities, is remarkable.

One hacker with dexterity issues was unable to pay for shopping because he could not hold his credit card firmly enough to insert it into a machine. He experimented with drilling a tiny hole in one corner of an old card, and looped a piece of fishing line through it. By putting his index finger through the loop, he was able to better grip the card between that finger and thumb, and therefore could pay for items. Another hacker found a way to integrate crutch holders into her bicycle so that when she finished cycling, she could seamlessly carry on walking.

While these may seem like small-scale hacks, they can grow quickly. Architect Betsey Farber, who had arthritis, came up with the idea for more comfortable kitchen utensils for her hands, with family members starting up the OXO utensils company in 1990 with fifteen items at a trade show. Sold for almost $300 million in 2004, the company philosophy still strongly echoes the disability lifehacker movement: "We notice things. We notice pain points and pains-in-the-neck . . . We see opportunities to improve a product or a process, or a part of everyday life, and we make things that make things better."

If you are reading this on a digital device, and swiping through the pages, you have an original lifehacker to thank. John Elias, an engineering PhD student in the 1990s, found that his severe repetitive stress syndrome, caused by carpal tunnel syndrome, interfered with his ability to study and work. Working with his adviser, Wayne Westerman, he developed prototype touch screen technologies and established a company called FingerWorks. Steve Jobs's Apple, Inc., bought Finger-Works in 2005, giving both Elias and Westerman senior engineering roles in the company. The advances they had made were used in the first iPhone touch screen, launched in 2007, which has since revolutionized how we consume digital information.

In their purposeful and creative approach to improving their lives, disability lifehackers like Farber and Elias have opened up new possibilities for the population at large. Key to the super-successful disability lifehacks has been how they open the door to entire populations of new users who were previously locked out of markets by making the benefits they provide easier, faster, or cheaper to access. Once they find themselves in a mass market, these products give more consumers access to the benefits originally designed for and by people with disabilities.

In some cases, as with the touch screen, disability lifehacks have created and transformed entire industries. In other cases, the impact is more niche but still significant. For example, people with disabilities campaigned hard for accessibility technologies such as captioning on TV programs, films, and social media that have since become ubiquitous. Studies show that making information more accessible and easier to read for partially sighted and blind people has also benefited people from ethnic minorities and people from older generations who might find it easier to read than to understand words verbally. My own experience of arguing with my mother about having subtitles on or off when we watch TV together is not unique.

Harnessing their creative capabilities more effectively means that the needs and opportunities of people living with disabilities can be better met. Given that in the US alone there are sixty-one million adults living with disabilities, this is not just about the considerable impact on quality of life, but also a significant market to be tapped into. Moreover, as several of the examples above illustrate, scaling up disability lifehacks can have a wider transformative effect. This happens materially, in terms of bringing those same benefits to the wider population. But greater involvement and engagement of people with disabilities in innovation efforts will also help to overcome the "them and us" gap that still exists across society, which sees those living with disabilities as different from and lesser than the norm and underpins the severe discrimination they continue to face around the world.

PURPOSE IS A DOING WORD

In 2005, I was working in refugee camps in southern Africa with a remarkable organization, Africa Humanitarian Action (AHA). In my experience of disaster response, there were many false heroes, but genuinely heroic organizations such as this one were often overlooked. AHA was a self-avowed "black NGO" formed in the midst of the brutality and horror of the Rwandan genocide of 1994. It started with a single medical team operating out of the Rwandan capital, Kigali. The organization assembled medical professionals from seven African countries to run health centers across Rwanda, becoming the first African-owned-and-operated NGO to operate during the genocide response.

AHA provided unprecedented twenty-four-hour emergency health services and regular out- and inpatient care. After the operation in Rwanda demonstrated its success, the organization grew to become a pan-African organization, working to provide medical aid to some of the most impoverished communities anywhere in the world.

What struck me at the time, and what has remained with me, is not just what they did; it is how they operated, across all the countries where I worked with them: Angola, the Democratic Republic of the Congo, Ethiopia, Liberia, Namibia, and Zambia. The organization was making what seemed to be impossible decisions on a daily basis: Do we provide food or water to this remote desert community in Namibia in the wake of a drought? How do we manage to keep our services going in Darfur when our budget is to be cut in half? How can we maintain a refugee camp while the local people living just beyond its fences are also starving and in need of assistance?

Every evening, instead of heading to the local bar like many foreign aid workers of my acquaintance, staff would sit together and talk about their day, their ups and downs, their fears and hopes, and they would share ideas about what they could do differently. Sometimes there would be tears, sometimes laughter; but no one ever finished a meeting feeling as worried as when they started, or indeed with the same ideas that they started with. We were working in extreme conditions and those around me coped by turning to novel, original, and entrepreneurial approaches. I documented these Upshifting efforts in a nascent fashion—although I didn't understand them fully at the time—and wrote a strategy for the organization to learn about and share these remarkable processes better across its countries of operation. I also shared these experiences more widely.

What really made AHA unique in retrospect, and perhaps hard to replicate, was that many of the staff members providing support to displaced communities were themselves refugees. It could not have been coincidental that these most tenacious of humanitarians were also people who had lived through crises themselves.

Much like the emergency health workers, leaders forged in crises, and disability innovators, refugees have an ability to overcome adversity. Evidence shows it is not down to some innate and inherited quality. A fascinating study of Palestinian refugees in Lebanon looked at their entrepreneurial efforts, undertaking detailed, in-depth interviews over a fifteen-month period. What it found was that the resilience of refugee entrepreneurs was not an "endowment" that they were gifted with but was created with time, effort, and purpose. The refugees with the greatest sense of resilience developed it thanks to their engagement in entrepreneurial actions, which broadened and deepened their social skills and integration both with other refugees and also the "host" Lebanese population.

When the components of refugee resilience were examined in more detail, they echoed the three qualities of leaders forged in crises:

1. Having a broad purpose and goal to achieve both moral and material gains and a sense of realistic optimism.
2. Proactive problem-solving behaviors based on self-reliance.
3. Multiple ways of framing and enacting collective action and a sense of belonging.

For the most successful, this purpose was not something they simply deployed—it is something they *built*. What's more, the very nature of entrepreneurial activity played a vital role in building their sense of purpose—but it is both a consequence and an outcome: there is a mutually beneficial relationship between the two.

This dynamic interaction between entrepreneurship and purpose has also been found by management scholars looking at entrepreneurs on the other side of the world. Much has been made of the idea that successful entrepreneurs are driven to greatness by the "fire of desire." Interviews with angel investors in Silicon Valley in California—high-net-worth individuals who invest in early-stage entrepreneurs—found that a sense of purpose, defined as the ability to sustain passion, action, and energy despite obstacles, was the quality that most inspired them to invest.

The same group of investors was presented with a series of hypothetical investment opportunities and asked to indicate the level of probability they would invest in each. Not only did the fictional entrepreneurs with greater levels of purpose have a greater chance of investment, it also turned out that the more entrepreneurial experience investors themselves had, the more they valued a sense of purpose in their potential investees. These investors "know that there are many uncertainties in the path to founding a viable venture, and that individuals who refuse to quit in the presence of adversity . . . are most likely to be able to overcome these uncertainties."

Importantly, as with the refugee entrepreneurs, this sense of purpose was not seen as a quality that entrepreneurs simply possessed. Rather purpose was an emotional and motivational process, "a doing word." Purpose was not something entrepreneurs had, but something they did; something they felt at a visceral level when engaging in activities that were relevant to their identity and goals.

In similar ways, the long conversations into the night were not a means of *being* purposeful for the teams at AHA. They were a way of *developing* purpose. And that sense of purpose and meaning fed back and strengthened our efforts. The same was true for the team at the Royal Victoria Hospital, and the disability lifehackers. And it has also proved true for those teams and organizations working to deal with the ongoing global pandemic.

It's there in the surgeons on the front line of the response, who are reporting not just sky-high levels of stress, but also unprecedentedly high levels of career satisfaction.

It is there in the hospitals that have found ways to build staff resilience and well-being through the prolonged crisis. A survey undertaken among frontline health workers in New York noted that despite the considerable emotional cost of the pandemic on their well-being, almost two-thirds of all participants reported a greater sense of purpose and meaning as a result of the pandemic.

Even among business start-ups that have been established in the worst economic climate in living memory, successes have been linked clearly to their sense of purpose.

This is yet more testament to the transformative power of purpose, even in the most extraordinary of circumstances. As Victor Frankl found, "life is never made unbearable by circumstances, but only by lack of meaning and purpose."

4

THE POWER OF UPSHIFT

A REPRISE

WHEN MOON SHOTS GO WRONG

How is it possible that a modest Chinese herbologist could have won the Nobel Prize for medicine in 2015 without a doctorate, without a reputation, and without any experience in a major national academy? The same way that the Apollo 13 mission successfully brought home the astronauts on that ill-fated lunar mission.

While one of these stories is famous to the point of mythology, the other is unknown to the point of obscurity. But both bear the hallmarks of Upshifting: mentality, originality, and purpose. In this closing chapter of part I, I want to bring together the three core ingredients of Upshifting as they were manifested in these two very different cases to show how they can be "the difference that makes a difference" to individuals and teams working in the most extraordinary of circumstances.

For one, both of these achievements were the result of research projects called for by the leaders of their respective countries in the context of the Cold War. President John F. Kennedy's 1960 moon shot is the most famous, not least because of its remarkable success. It has become shorthand for any attempt to solve big, hard problems that demand significant investments of time and money, along with innovative technology and thinking. The same era saw another moon shot, more secretive but just as political. In 1967, at the height of the war in Vietnam, China's chairman, Mao Zedong, issued an order to find a cure for malaria following a personal request from the Vietnamese premier, Ho Chi Minh, whose Vietcong troops were being decimated by the disease.

Fifty-six hours into the third manned venture to the moon's surface, NASA's Apollo 13 mission, an explosion in an oxygen tank took out an entire half of the spacecraft, and left the crew without fuel for propulsion or life support. The lunar module had to be repurposed mid-mission to support the crew for three nerve-wracking days. Tens of millions of people watched the splashdown of the astronauts in the South Pacific Ocean. But the popular understanding of how this happened, shaped by Hollywood and our own collective tendency for mythmaking, is largely inaccurate. My research, which includes analysis of in-depth interviews with the original crew, reveals the biggest misunderstandings of the rescue and gives a more nuanced explanation about what underpinned the successful return of the three astronauts.

I want to contrast this retelling of what actually led to the successful rescue of the three astronauts with the story of how a malaria cure was found, against all the odds, by Chinese herbologist Tu Youyou. Three years before the drama of Apollo 13, Chairman Mao issued a secret order to find a cure for malaria, to keep the Vietcong "combat ready." Vietnamese soldiers had a saying: "We're not afraid of American imperialists, but we are afraid of malaria." And the "imperialists" were far from immune. Three years earlier, malaria-related casualties among the US forces were five times greater than casualties from direct combat. In 1965, nearly half of all US military personnel—some eight hundred thousand troops—were infected. As a result, tackling malaria had become a top military priority for both sides.

The US military threw its considerable resources at the problem. Thousands of scientists, led by Walter Reed National Military Medical Center, were looking at all the possible biomedical treatments, screening almost a quarter million drugs. In China, the project was launched with much less fanfare and resources. Some six hundred scientists came together on May 23, 1967, with the date giving Project 523 its code name. For many in the Chinese government the project was so secret it was only ever known as "523." Although none of the participants knew it at the time, the project would become something of a shelter; a malarial ark against the persecution of the Cultural Revolution, which saw scientists and researchers come under symbolic and actual attacks.

So, on the one hand, a team was trying to survive but also not fail on multiple levels; technological, human, organizational, and political. On the other, there was an attempt to find a cure for malaria, one of the oldest and most prevalent diseases in human history, in one of the most endemic regions in the world, during a brutal war. We will see how these two stories about performance under pressure can help us to understand the basic principles of successful Upshifting. It will also, I hope, demonstrate the power of the Upshifting concepts to unpack and get to the essence of these remarkable efforts.

MENTALITY

We learned in chapter 1 about getting ourselves into a challenge mentality as a way of getting past the pressure of the moment, and transcending the limitations we might face. In the *Apollo 13* movie, Ed Harris, who plays flight director Gene Kranz, says to a group that has spent the previous few minutes arguing about the impossibility of the task: "We've never lost an American in space; we're sure as hell not going to lose one on my watch. Failure is not an option." His words render the team silent as they contemplate the enormity of their task. It became the tagline for the movie and even became the name of the real-life Kranz's autobiography.

But it never actually happened. Kranz's actual words, and the reaction to them, were completely different. He relayed them from memory fifty years later, and I was able to confirm these with reference to the Mission Control records, as follows: "This crew is coming home. You have to believe it. Your people have to believe it. And we must make it happen." It may seem like a subtle point, but the real Kranz's words got the team into a challenge state. By contrast, the fictional words are more indicative of a threat state. In fact, the team went further: they did not spend any time contemplating disaster. The second-shift flight director, Glynn Lunney (who did not feature in the movie), reinforced this later: "If you spend your time thinking about the crew dying, you're only going to make that eventuality more likely."

Focusing on not failing is not the same as focusing on success. Rather than leaving the team silent, the real speech electrified the room. "Everybody started talking and throwing ideas around," one of Kranz's three lieutenants recalled. Interviews with the team give us even more detail about the process the entire team went through. As Fred Haise put it: "I never felt we were in a hopeless [situation] . . . We never had that emotion at all." Flight director Gerald Griffin had a different take on things: "Some years later I went back to the log and looked up that mission. My writing was almost illegible, I was so damned nervous. And I remember the exhilaration running through me."

The Project 523 malaria team faced rather different constraints from the outset. Because their work was considered a military secret, no communication with the outside world about their research was allowed; in any case, during the tumult of the Cultural Revolution, publication in scientific journals was forbidden. For these reasons, no one outside of Project 523 knew about the work. It may as well not have existed. But within the project, the atmosphere was energized. As noted in the official report, coauthored by four of the senior leaders: "Project 523 became a great spiritual force to unify all participating and collaborative research teams."

The US effort led by Walter Reed had many biomedical researchers working together. In four years, they had screened 214,000 chemicals but without any success. Project 523, with their meager resources, could not hope to compete with the sheer human and financial capital of the Walter Reed effort. But their approach was completely different.

The original project document, translated from Chinese, sets out the need for Project 523 to: "integrate far and near, integrate Chinese and Western medicines . . . emphasize innovation, unify plans, divide labor to work together."

Project 523 had two wings: A network of biomedical researchers and Tu Youyou's group, which was tasked with examining traditional Chinese medicines. When she started her search for a malaria treatment, Tu knew that the Walter Reed Institute had already undertaken extensive testing without success. By comparison with the US effort, at that time China lacked scientific professionals, had out-of-date equipment, and had no serious experience in drug research. For many, this may have been a dismal prospect. Tu would later describe her work as taking place in "under-resourced research conditions," a comment that would come to be seen as characteristic of her understated nature.

But Tu took heart from the quasi-mythical founder of Chinese herbal medicine, believed to have lived around 3,000 BCE. Legend tells that Shennong, also known as the "divine farmer" in China, developed and taught his people essential agricultural practices and the use of medicinal plants.

All students of traditional Chinese medicine know that it begins with the the fabled story of how Shennong ingested over a hundred different types of herbs himself to understand their efficacy and toxicity. For Tu

Youyou, Shennong proved a source of motivation and courage. "Shennong tasted a hundred herbs, why couldn't we?"

The common lesson about mentality from Apollo 13 and Project 523 is about how far we can advance when we move to a state of seeing problems as challenges rather than threats, individually and collectively.

The way we choose to see the world and problems we face lays a foundation for the way we respond to pressure, stress, and crisis—and this enables Upshifting. Although these examples are clearly world-changing in their scope, as we have seen throughout the chapters so far, the same principles play out in more everyday settings.

ORIGINALITY

Jim Lovell, played in the *Apollo 13* movie by Tom Hanks—who seems to be something of a typecast Upshifter—said, "We were given the situation to really exercise our skills, and our talents to take a situation which was almost certainly catastrophic, and come home safely. . . . Of all the flights— including 11 . . . 13 exemplified a *real test pilot's flight*" (emphasis added).

Test pilots work to fly and undertake real-time evaluations of experimental, newly produced, and modified aircraft, with specific maneuvers. The term "flight envelope" was one of the earliest test pilot phrases, meaning the range of speeds and internal and external conditions at which an aircraft can safely operate. Test pilots helped build up a picture of the flight envelope for new and existing aircraft, giving rise to the phrase "pushing the envelope" to describe when they were pushing a craft to new limits. In fact, this phrase moved into popular vernacular thanks to Tom Wolfe's book *The Right Stuff*, which chronicled American pilots who tested high-speed aircraft, including early astronauts.

Over the course of the eighty-seven hours of the crisis, pushing the envelope took several teams and leaders working in shifts. It wasn't a sprint *or* a marathon: it was a relay. The NASA chief later said there was "a true spirit of teamwork—the ability to know when your part is done, when someone new can bring something better than you can." In a sense, divergent thinking was built into the mix from the outset. As one crew member put it: "We never were with our backs to the wall, where there was no more ideas, or nothing else to try, or no possible solution. That never came." Indeed, despite the famous depiction of crisis-driven improvisation that many people associate with the recovery mission, much of what was done during that eighty-seven hours had been tested and simulated prior to the accident.

This richness of original thinking was not just because many people were involved. There was also a diversity of skills and styles on show. In the *Apollo 13* movie, both Kranz in Mission Control in Houston and Jim Lovell on the spacecraft are similar kinds of leaders, who quietly bond over the course of the movie thanks to their shared qualities: rationality, motivational skills, and the ability to drive for results. But actually, there were two types of leaders that contributed to success. Kranz was a technical expert on spacecraft. He was the best person to be in charge of the initial phase, which saw second-by-second equipment failure, and the step-by-step reconfigurations necessary to maintain the systems. Lunney, on the other hand, was a flight dynamics specialist. According to the NASA boss, Kranz was there at the right time to make the decisions that had to be made rapidly; and then, when Lunney took over, the latter brought a calmness to the control center to do the right things once the situation had become more stable.

Not only was Kranz and Lunney's partnership a significant illustration of the importance of diversity, but so too was Lunney's specific leadership approach. Interestingly, these are not recounted in the movie or included in any popular retellings.

Ken Mattingly—a member of the Mission Control team who would have been on board Apollo 13 if not for a failed medical exam—gave the following pertinent account of how Lunney's leadership enabled the team to embrace the novel problem they faced and help meet it with divergent approaches.

In the midst of this, they did the hand-over. Glynn [Lunney] stepped in, and this is where the most magnificent display of personal leadership that I've ever seen, because there was confusion, not chaos. People were confused. They were highly trained to do things, but this was out of the experience base, and it was real, and we didn't understand it. Glynn came in, and Gene was still there, so I mean nobody left. We had these extra people, and they got into a discussion. Glynn stood up and in his quiet way—the contrast between Glynn's speech and Gene Kranz's speech is really stark. Gene's crisp and precise and sometimes loud, and Glynn is quiet and laid back. He went around and he just started asking people, and my sense was that he was asking questions that were relevant, but not particularly important, but he went to every position to the room and gave them a question to get back to him on. So, all of a sudden . . . my sense was, it didn't matter what question he asked. It was just getting your mind on something constructive, and then it'll all take care of itself. You could almost feel the room settle down. Emotions didn't go away and all that, but all of a sudden, they were back to the team that had trained with a sharp focus.

Let's turn back to the malaria work. This also saw a kind of relay race, as in the Apollo mission. It was manifested through the exchange of ideas and sharing of tasks between the two wings. There was a clear sense that the research teams in either wing had distinct roles but needed regular communication. The first team—the biomedical team—scanned possible drugs, plants, and herbs, worked on chemical isolation of specific active ingredients in plants, and synthesized them for trials. Tu's team was tasked with two jobs: screening ancient medical texts and identifying traditional treatments that were already in use in communities across China, often closely guarded.

They analyzed more than two thousand Chinese herbal preparations, of which a third had potentially antimalarial properties. The results were shared back and forth between the teams to ensure continuous learning. Extracts were taken and tested, and information fed back to inform more primary research. Progress was made thanks to this highly collaborative continuous exchange between usually separate domains of medical knowledge. Indeed, it was this collaboration that was eventually seen as underpinning the ability of Project 523 to turn the disadvantages it faced relative to the US effort into an advantage.

Tu and her team scoured ancient Chinese manuscripts for leads on substances that might help them defeat malaria. In a recently excavated nine-hundred-year-old text titled *The Manual of Clinical Practice and Emergency Remedies* they found a mention of sweet wormwood— *Artemisia annua* or *qing hao*—being used to treat malaria. This was backed up by findings from community research in Nanjing district, which found that the same plant was used to prevent and treat malaria, and even had local sayings associated with its use.

Although initial tests were promising, attempts by botanists to replicate the results by boiling the herb and using the resulting mixture had failed. Tu returned to the texts. She read the following: "A handful of *qing hao* immersed with 2 liters of water, wring out the juice and drink it all." She realized that the means of extraction used by the botanists, which involved boiling the plant, could be the source of the problem. Sure enough, a lower-temperature extraction process, where instead of water Tu used ethanol because of its lower boiling point of 78°C (173°F), led to a solution from which the antimalarial results were significant. The active ingredient would become known as artemisinin.

Once artemisinin had been identified, the collaboration between the two wings played a vital role in ensuring the process of testing and synthesis could overcome constraints through creative adaptations. This is from the official project report:

> [Tu's] institute had a good start in extracting *qinghaosu*, but then ran into difficulty . . . Yunnan Institute of Materia Medica started late, but had a smooth run, picked up, and raced forward . . . Without the latter two units' participation making such good progress, *qinghaosu* research might have been terminated prematurely or delayed for at least a few years.

This shows how the next step for Upshifting is to think and act in original ways, to see and act anew. Not to rely on established understanding and approaches, as the US effort did, but—like the most creative of architects—to "have sport with what [they] know" of the world, and to be fearless in our curiosity. Changing how we think is not easy. But when we do it under pressure, it not only provides a means of tackling the novel situations we face, but it also helps to build our inner resilience and confidence in the face of that pressure.

PURPOSE

The Hollywood narrative of Mission Control's work during the Apollo 13 rescue was that heroic leadership was paramount, and an elite team led by Flight Director Gene Kranz made all the difference ("Failure is not an option").

In the face of such a high-stakes scenario, it would indeed have been tempting to impose a command-and-control way of working.

In reality, Kranz worked hard to empower his most junior team members, giving them total ownership of their specialist stations.

He said later of his leadership approach: "It was a question of convincing the people that we were smart enough, sharp enough, fast enough, that as a team we could take an impossible situation and recover from it." The delegation of authority and deference to expertise were essential: Mission Control would interrogate the teams' recommendations, but not second-guess them. Ken Mattingly commented:

> There was no room for any distraction, no room for politics. There's no personalities getting in. I don't care who's got the right answer, just get it right, and it's okay. It didn't matter if it's the new kid on the block or the guy who's retired. Anyone who's got an answer to our problems is sought after and appreciated. And you don't get to work in that climate very often.

The malaria search demonstrated a similar tenacity, although it was a slower burn. As noted in the official Project 523 report, coauthored by four of the senior leaders:

> [There was] a sense of responsibility for everyone, so they brought to the work not only intelligence, knowledge and skills, but also their passion. From laboratories in cities to clinical trials in the countryside and in the mountainous areas, everyone responded in spite of many difficulties, to achieve the stated goal.

By 1969, the Project 523 team had looked at thousands of possible drugs with no success. The Walter Reed work also failed to produce significant results. It was at this point that the government brought in the Academy of Traditional Chinese Medicine to look for possible cures. These were the circumstances that led Tu Youyou to be appointed overall director of Project 523.

Tu's work on Project 523 started with a four-month trip to Hainan, a tropical island at the southernmost point of China. Tu was deeply affected by the plight of children. As she subsequently put it, "I saw a lot of children who were in the late stages of malaria. Those kids died very quickly." As can sometimes happen, this heightened empathy for those most affected by the malaria epidemic came at a cost to Tu's own family. Her husband had been exiled and she was the sole carer for their daughter, who she had to leave in a local nursery for six months. When she returned from Hainan, her daughter didn't recognize her and refused to leave the nursery to go home. As Tu said later, "The work was the top priority, so I was certainly willing to sacrifice my personal life." It would be three years before she saw her family again.

Tu's tenacity was shared by the lead researcher involved in the chemical synthesis of new drugs, Zhou Yiqing. He had firsthand experience of the battlefield, traveling the Ho Chi Minh Trail with soldiers as bombs showered the rain forest. Zhou recalled that the thing that bothered him most was malaria-ridden soldiers "begging me to save their lives . . . I just could not help them." The experiences of the researchers were vital for instilling in them the belief that they not only could find a treatment, but that they would.

But even when artemisinin had been identified and found to have a positive effect, problems remained. Most pharmaceutical laboratories were closed because of the Cultural Revolution. There was no manufacturing support. Tu's team had to extract the active ingredients themselves at home using household vats, pots, and pans. The lack of proper equipment and ventilation meant that they started to exhibit symptoms of ill health. But they persevered. Tu recalled this time as being a "very laborious and tedious job, in particular when you faced one failure and another. This was the most challenging stage of the project."

They finally faced a crunch point as dramatic as any on the Apollo 13 program: the toxicity of artemisinin for human subjects was still in question. They had just a few short weeks left before the end of the malaria season, and if they could not trial artemisinin safely and ethically in that period, they would have to wait another whole year. Tu and two colleagues wrote and asked for permission from their institutions to take artemisinin extracts themselves under close supervision for a week. After no ill effects, the drug was tested on malaria-infected laborers in Hainan. Their symptoms disappeared within two days. This was enough to start a concentrated effort to trial and synthesize artemisinin. As Tu later put it: "Our desire [to] have the medicine for our patients as soon as possible was the real driving force behind our action."

Both situations show how common purpose, persistence, and resilience can be generated from novel and high-pressure situations. As we saw in chapter 3, of all the events that can deeply engage people in their jobs, the most important are those that give people the opportunity to make progress in meaningful work. When constraints and challenges help us to find meaning in our work, they can also instill a sense of ownership. It means something to us personally. We become more committed, more intrinsically motivated, and more engaged. This is precisely what happened in the Apollo 13 mission and Project 523.

"NOT MUCH OF A SURPRISE"

The story of Apollo 13 success is so famous that it has become riddled with myths and inaccuracies with each retelling. Tu Youyou's story deserves to be accurately retold as often and to be as well known.

For sheer life-saving impact, though, there is no comparison. The outcome of the Apollo rescue mission saved three lives—and probably many more careers in the immediate aftermath. And the lessons had a knock-on impact on many subsequent space programs. In 2018, *214 million* artemisinin-based treatment courses for malaria were provided to people around the world. Today, thanks to Tu's discovery, the chances of dying from a severe case of malaria have halved from one in five to one in ten. According to the London School of Hygiene & Tropical Medicine, the total number of deaths has been reduced by over 75 percent. In absolute numbers, the World Health Organization estimates this is a drop from two million deaths globally to around 435,000 in 2017. Tu is one of the small number of people in human history—perhaps numbering fewer than ten people in total—who can reasonably claim to have saved over a million human lives.

Awarding Tu its Clinical Medical Research Award in 2011, the Lasker Foundation, hailed her discovery as "the most important pharmaceutical intervention in the last half-century." In 2015, Tu won the Nobel Prize in Physiology or Medicine. She was the first mainland Chinese scientist to have received a Nobel Prize in a scientific category, and she did so without a doctorate, medical degree, or any experience outside of China.

In a rare moment of departure from her usual almost excruciating levels of modesty, she remarked: "It was a bit of a surprise, but not much of one."

These three ingredients—mentality, originality, and purpose—help make sense of transformative performance under pressure. There is an interplay between these three ingredients: a virtuous Upshifting circle.

We see it in the example of the Royal Victoria, but not in the other Belfast hospitals.

We see traces of it in the remarkable stories of early human survival and creativity in Cape Floral, but not in other early human groups who didn't thrive in the same way.

We see it in some entrepreneurs and how they perform over time—but not in others.

And we can see it in our own lives, both in comparison to others, and across our diverse experiences.

Upshifting is not a given. It is a repeated set of behaviors that, over time, become engrained habits, ways of thinking and relating, and, ultimately, a way of being.

So far in *Upshift*, you have been given the map, been invited to study the terrain, and learned about some of the travelers who have been here before.

Now you are about to learn what happens when the rubber hits the road.

PART II

THE ART
OF UPSHIFT

INTRODUCTION

In part I, we saw that people have been Upshifting for as long as humanity has existed. We have always found ways of overcoming and navigating stress and pressure. Moments of crisis have inspired us to transcend our limitations and to overcome the inertia of habits, the weight of accepted practices, and the thick treacle of institutional memory. Indeed, as we have seen, archaeologists increasingly believe this is not just a defining feature of our species, but a formative one. Using stress and pressure as catalysts for growth and transformation is part of what made us—and continues to make us—human.

Yet today, despite the wealth of evidence and examples, many of us live in fear of stress and pressure. We seek not to grow out of the shadows they cast, but instead see them as a threat, a source of dread and anxiety. This is not just counterproductive; it's downright unhealthy. But we can and should change our way of thinking about stress and pressure, building on the adage "no pressure, no diamonds." I have set out the science behind Upshifting, drawing on neuroscience, physiology, individual psychology, and group behavior. I explained the three critical ingredients of *mentality*, *originality*, and *purpose* that have enabled Upshifters to mine diamonds from the depths of crises.

The experiences of Keith Jarrett, Captain Chesley "Sully"

Sullenberger, the Royal Victoria Hospital, Belfast, and others illustrated in depth each of the ingredients of Upshifting, and the science and research that underpin these. The examples of the Apollo 13 rescue and Project 523's discovery of a cure for malaria showed us how the three ingredients work in combination.

What we also saw in part I is that people Upshift in different ways. Rather than a singular approach, Upshifting builds on our individual and distinctive problem-solving skills and styles. Katarina Linczeny-iova spends countless hours imagining and refining each of her three-minute dives, so that as soon as she clicks on that master switch, her Upshift is underway. Tu Youyou fused together traditional and modern medicine to (re)discover an ancient cure for malaria. Sister Kate O'Hanlon's passion and values allowed her to orchestrate and channel the inventiveness and skills of those around her to solve seemingly impossible tasks during a protracted conflict.

These represent just three of the six Upshift archetypes that we will explore in part II.

The first archetype is that of Challengers, preferred by those of us who are skilled in constructively disrupting the status quo around us. Protest leaders such as Rosa Parks and Greta Thunberg are classic Challengers, but so are those who actively look for new working or teaching practices in their everyday jobs and relationships.

The second archetype relates to how Crafters understand and experiment with social, physical, and technological processes. Thomas Edison's ten thousand iterations of the filament light bulb before his final success is an iconic example. Crafters are not merely creative thinkers: they are good creative *tinkerers*, constantly tweaking, prodding, poking, breaking, rebuilding, testing, and improving. Crafters see the world as their laboratory.

The third archetype relates to how Combiners fuse ideas from disparate fields and endeavors. The Wright brothers' "discovery" of flight is a beautiful illustration of the combining mentality, weaving together insights from observing birds in flight with innovations from the bicycle industry. Some of the most important Upshifts happen at the intersection of different fields, disciplines, and departments, and Combiners are people who can straddle them all with apparent ease.

The fourth archetype shows how Connectors create bridges and networks between different kinds of people with no sense of social hierarchy. Connectors get their energy from meeting new people, and by bridging and brokering gaps in their social networks. Whenever you say, "It's a small world," you likely have Connectors to thank. Paul Revere, a pivotal figure in the American Revolution, was a famous Connector.

The fifth archetype concerns the role that Corroborators play in proving that new ideas work—or don't. It was a hunger for data and a passion for analysis that enabled the gifted mathematician Katherine Johnson to calculate the trajectory of the first US-crewed Earth orbit, the moon landings, and the return path for the Apollo 13 crew—a phenomenal responsibility, and one that relied on her rigor in testing and retesting the speculations of NASA's aeronautical engineers. Corroborators push for logical and critical thinking, both within themselves and from others.

The sixth and final archetype shows how Conductors orchestrate different minds to bring about change. Abraham Lincoln demonstrated skills as a Conductor throughout his career, using his unique abilities not just to drive change, but to facilitate it. Constructing his first cabinet as the original "team of rivals"—smart people with different viewpoints who would challenge one another and him— was a classic Conductor maneuver. Sports managers and even school coaches often do the same to create winning teams. Conductors bring together people with different skills and personalities to achieve a common goal.

It was in the fog and the wild of humanitarian responses to wars and disasters that I first observed and documented these diverse styles.

Some people I worked with just seemed to thrive under the pressure of crises, while others didn't. But as I observed and worked on disasters with an ever-changing cast of characters around me, I learned that many of these styles were not in fact predetermined. With practice, and by leaning into our natural tendencies that match a particular one or more of the six archetypes, we can all become better Upshifters.

Much like the leaders throughout history who were "forged in crisis," my colleagues developed certain ways of working as a result of the pressure and stress we were facing. Over time, these became habits, repeated patterns.

I learned about and documented what I call the art of Upshift— the six archetypal ways in which people perform under pressure— some time before I had the chance to dig deeper and identify the scientific foundations.

In the second part of this book, then, my purpose is twofold. First, I want to tell these stories of how different kinds of people Upshift, and show not just what they do but how it feels, and outline some tips and tricks so we can try out different appproaches.

I do this to help us understand and internalize the art of Upshift: to find ourelves in these archetypes, to think about which ones we might instinctively prefer or habitually turn to. In doing so my hope is that we will be able to recognize and build on our individual strengths, and perhaps address some of our blind spots and biases in the process.

Second, in exploring these archetypes and how they have manifested in every sphere of human endeavor and throughout history, I want to open up possibilities beyond the individual level as to how the ideas of Upshift can help transform how we think about and solve problems in our lives, our work, in society, and in the world at large.

5

CHALLENGERS

CHANGE THE RULES

"THERE HAS TO BE A BETTER WAY"

One of the most famous quotes about innovation is often attributed to Henry Ford, who reputedly said, "If I had asked people what they wanted, they would have said a faster horse." This phrase has entered the annals of business thinking as a signal of Ford's dismissiveness of customers and users, and the need to rely instead on visionary leaders who disrupt the status quo.

It is rather ironic, therefore, that one of the most disruptive innovations in international crisis response in the last fifty years came about because its instigator decided to ride a horse rather than drive a car.

In the summer of 1985, a young Irish medical student named Steve Collins had managed to convince his medical school to give him a year off to do a public-health placement in Uganda. Steve Collins was spending his summer traveling across Africa. He had reached the western borders of Sudan, the region of Darfur, where a full-scale famine had just broken out. Some years later he said of his experience:

"I realised you couldn't be a tourist in a famine, and so I turned up at a refugee camp, and volunteered. [They] needed somebody to survey the villages in the surrounding desert, and I said I'd do it, on foot. The first day, I walked 15 miles across the desert to the nearest village."

At every house he visited, Collins was given some *merisa*, a local sour beer made from fermented sorghum grain and dates that the people had been given in excess. After a few hours, drunk and stumbling in the baking heat, Collins passed out. This is far from an ideal situation in any work context, but when your working environment is on the fringes of the Sahara Desert, it is a life-threatening mistake. By a stroke of luck, he was found by his new boss, who suggested a horse would be a better way of traversing the desert.

Instead of going on to Uganda as he had originally planned, Collins spent nine months riding around Darfur villages, trying as best he could to support the local communities through the challenging dry season. As he later put it: "In Sudan, I had realised that nutrition was the basis of everything. If you don't have the right nutrition in the first two years of life, the brain doesn't develop properly, and you can't learn." But more than that, he also learned that nutrition was very different from what was commonly understood. Collins learned this not because of experienced colleagues or informative textbooks, but thanks to the fact he was on horseback.

When I said goodbye to my grandmother during the throes of the Sri Lankan civil war, I told her I would return to get her when I was eighteen in a white Land Rover. Anyone who has ever had any interaction with aid workers will know how strong the association is with these vehicles. They are more than just a form of transportation—they are a symbol of money and authority, a symbol that Collins had foregone by riding a horse.

As he told me from his farm overlooking the hills outside Cork in western Ireland: "Wearing robes and a hat, on a horse, I slipped in and could talk to anyone, and got a very different take on things." What he learned was that the providers of aid had their own strongly held views on the famine, the problems being faced, and the solutions. "The thing that really stood out was that no matter how expert we thought we were, we could only improve nutrition by looking at the problem from the community perspective." The reality, though, was that much of the aid that was being provided was being imposed.

Returning to finish his medical degree, Collins was posted in Jamaica for his internship. The hospital was next door to the Tropical Metabolism Research Unit (TMRU), an organization established in the 1950s when under-five mortality and malnutrition rates in the West Indies were as high as in parts of Sub-Saharan Africa. Collins made some valuable connections and leavened his field-based understanding with more of the science of nutrition. A few years later, in 1992, Collins volunteered with the Irish NGO Concern, and was sent to Somalia, on the front line of the worst famine on record. Unlike the Ethiopian famine, which also made the headlines in the 1980s, not just children but also adults were dying, and no one knew why. Some of us will remember the news reports and the shocking images of emaciated people, like walking skeletons covered in skin.

Collins was put in charge of a specialist hospital for treating adults with severe malnutrition. This was in fact the first time that such a facility had been established anywhere in the world in almost forty years: since the Second World War and the liberation of the death camps at Bergen-Belsen and Auschwitz. And with just cause. The people Collins was treating were 20 percent thinner than the prisoners liberated from Bergen-Belsen, with the lowest body mass index (a measure of body fat) ever recorded in living human beings.

It may seem strange now, but 1992 was not a time when computers were a standard part of most jobs. But thanks to a suggestion from a friend, Collins arrived in Somalia with a laptop. He was struck by the fact that information about what was happening was almost non-existent: "There was no data being gathered by the program staff in the feeding center because they were too busy supplying services. This struck me as a major issue limiting progress, so I began collecting data on all the adults we treated."

Armed with technology and insights from colleagues based at the TMRU in Jamaica, Collins captured and analyzed data, and used this to change the diets of the patients. He ended up having to order bags of nutrients and mix pastes and formulations himself. It worked. Collins's homegrown data-driven approach cut the death rate from three out of every four people to one in five—still tragic, but radically improved.

His achievements did not go unnoticed. In his subsequent publication in the prestigious journal *Nature*, Collins showed that those patients who had recovered demonstrated a new and dramatic level of human adaptation to starvation, where their BMI had dropped below ten kilograms per cubic meter. He would get a national honor from Queen Elizabeth II for this work. Not yet thirty years old, he was called upon to set up similar treatment centers in many other parts of the world experiencing famine. Suddenly, he was an expert. As he put it: "The reality was that I knew *almost* nothing, but everyone else knew *absolutely* nothing."

Then, within an already spiraling crisis, tragedy struck. During the brutal Liberian civil war in 1996, Collins set up feeding centers that attracted many families desperate for help. Unable to treat everyone who needed help as quickly as was needed, queues grew, and eventually an informal settlement formed around Collins's center. Although Collins had done the requisite water quality tests when he set up the hospital, the ramshackle aid shantytown that emerged around it had no facilities for keeping people safe and hygienic. A cholera outbreak struck, with people camping around the hospital starting to get sick and die. Then a member of staff at the hospital got sick and started transmitting cholera to the patients. Many more people died. One little girl became so ill due to receiving infected therapeutic milk that a desperate Collins ignored the vomit and risk of infection to himself and gave her mouth-to-mouth resuscitation. She died anyway.

Collins was understandably devastated and guilt ridden. He spent months in recovery, initially depressed, wracked by guilt and experiencing PTSD. As he told me, "I was meant to be an expert and I had screwed up." He went back to Jamaica, bought an old pirate boat, and spent months renovating it and living on it. The R&R was much needed, but there was something Collins could not stop saying to himself: "There has to be a better way." There was, and we shall see what it was later on in this chapter. But to get there, Collins had to upend half a century of nutritional knowledge, go up against the leading aid agencies of the day, and question his own source of expertise. He had to build on his early experiences and lessons from Sudan and use them to become a *Challenger*.

"TRANSFORMATIONAL CREATIVES"

We already learned in part I about the importance of mentality in keeping our cognitive bandwidth open when experiencing stress and pressure. We saw that Upshifting consists of a unique combination of characteristics and behaviors: a *mentality* that reframes stress and pressure, an openness to *originality* and novel approaches, and the pursuit of goals with a heightened sense of *purpose*. Less stress management and more "stress capitalization." I am convinced that Upshifting is a pathway and a possibility for all of us. But it is also the road less traveled. Part of the reason is that in the face of stress, pressure, and crises, most of us tend to double down on what we already know.

This is in part because of the physics and biology of our brains and how they underpin psychological biases. Our brains use about twenty watts of power on average—about the same as an average light bulb. While this might not seem like much, it is still 20 percent of our overall energy footprint for an organ that is only about 2 percent of our body weight. This disproportional allocation doesn't mean the brain is greedy. Far from it: our brains are constantly trying to find ways to use less energy and to be more efficient. One of the main ways this is done is through repetition.

It turns out that almost half of our behaviors are repeated from the past, with little attention paid to whether they are ideal or desirable. We use all manner of behavioral shortcuts in the name of cognitive efficiency. Sometimes these are helpful, and are known as heuristics. But other times they are habitual and are better described as traps.

This biological reality is reinforced by society. We are, of course, much more than neurons firing across a squidgy brain mass—but it turns out that institutions and norms are also resistant to novelty. Not only is it less cognitively efficient to go against the norm, it is also less socially acceptable: "The typical human tendency is to strive for consistency and status quo rather than to search for and enable new behaviours."

By their very nature, social groups need to collaborate, and this means working on the basis of predictability and consistency, reducing risk and minimizing error. While this is of course important, new ideas can often run in the exact opposite direction to this imperative: they are unpredictable, risky, and prone to failure. This is especially true in the heightened drama of crises where many of us insist on "doing the wrong thing righter." We may well have worked out that the problem we face is not a nail. But if the metaphorical hammer is in our hand, it is often easier and more convenient to just keep whacking and hoping for the best. Both cognitive efficiency and social acceptability are traps that we back into, protesting our ignorance, but from which we find it very hard to escape.

Which is why Challengers are so vital to the whole Upshifting enterprise. If we think back to the patterns of brain activity during Upshifting, we learned about how the software of our mind can influence the hardware of our brain chemistry. We also learned about the three brain networks—the imagination network that underpins novelty and empathy, the salience network that aids sensemaking and willpower, and the executive network that drives focus and decisions—and how they work together when we are thinking divergently. It turns out that both of these things work in tandem in the brains of Challengers. Gregory Berns, a neuroscientist, has done the most in-depth analysis of the brains of Challengers in different fields, from visual arts to science. He found that Challengers, above all else, were characterized by an ability to see or visualize alternatives, and a low level of social fear—or willingness to forgo acceptance. Challengers are people who have learned to navigate out of the twin traps of efficiency and acceptability.

When Challengers hit these walls, they don't just give up; they look for other ways forward. Upshifting combines the mentality of overcoming stress and pressure with the originality to look for alternative ways of thinking and acting. In this sense, Challengers are the catalysts of the whole Upshifting process. They are the *iconoclasts*.

Historically, this literally meant those people willing to torch the icons of the past to envisage new possibilities. These icons could be physical—like the "false idols" or "graven images" that were burned and broken in various passages of the Old Testament. But they could also be conceptual or symbolic. Maggie Boden, a colleague of mine at the University of Sussex and an influential cognitive psychologist, has argued that Challengers—or "transformational creatives," in her terms—are unafraid of questioning the rules of the game. Indeed, they thrive on it. They ask why, repeatedly, and do not follow the path of least resistance with the answers. This can mean adapting existing rules, dropping old rules, or adding new ones.

But what exactly is it about being under stress and pressure that pushes Challengers to do their rule-breaking, and what can the rest of us learn from them? To find out, I want to turn not to the annals of creativity but to an activity as far removed from innovation as most of us can imagine: our daily commute.

"WHEN WAS THE LAST TIME YOU DID SOMETHING FOR THE FIRST TIME?"

If you live in or near some form of metropolis, like more than half of the people on the planet, chances are that you will have some kind of travel card. Gone are the days when this was a plasticized card with a handwritten code. Today, many travel cards are digital, allowing for contactless travel across cities and in some places entire countries. This is not only more efficient, but it has created huge accumulations of data and made possible new types and levels of analysis. Every single touch in and out of the transportation network is a data point, every two points are a journey, and, as a whole, we have a daily treasure trove that can be mined for its secrets about human behavior. You may think that travel data is not that revealing, but there are features of smart card data that make it uniquely valuable: it is large, it is continuous, and it is comprehensive.

Armed with the right questions and tools, we can learn a huge amount about how and when people use transport: patterns of movement, preferred origins and destinations, and the behaviors of different kinds of travelers from daily-drudge commuters to infrequent users. Most relevant for us, this big data set contains real insights into human habits and choices, and the things that shape them. The journalist Tim Harford first alerted me to the potential of this phenomena in his account of University of Oxford research from 2016, which shows how commuter habits change in the face of unexpected events.

This was a remarkable study for its size and the precision of its findings. The data looks at how commuters responded to two days of significant labor strikes in February 2014 on the London Underground. On these days, many Tube stations were closed, with 60 percent of commuters unable to enter the network at their normal stations. Only about 50 percent could exit where they normally would. This meant that a large proportion of commuters had to experiment and explore new routes. What the Oxford researchers were keen to understand was the impact of this "forced experimentation."

They were especially interested to learn how the strikes changed peoples' behaviors in terms of lasting changes beyond the strike day: "When all stations were open again on February 7, did people switch back to their original paths, or did some of them stick to the alternative route that they found during the disruption?" This is a pertinent question, because commuters themselves typically view their travel experiences as highly negative, with many linking it to their overall experience of life satisfaction. Because of this, the assumption might be that actually London Underground users had little to improve: their commute was already pretty well optimized so as to minimize the negative impacts. The reality was quite different.

For a good proportion of commuters, the strikes were the start of doing more novel and interesting and useful things. Of the total, almost 6 percent found a better route to work. These adaptations meant that as a whole on those strike days, people spent only a fraction more time in transit: a journey that would have taken thirty minutes on a normal day would, on average, have taken thirty-two minutes on the strike days.

Even more interesting was what the researchers found when they looked at the overall impact of the strikes on the network. They also found that the negative cost to the whole of London for that period of the strike was exceeded by the positive cost of the savings caused by disruption of the strike. This is down to these experimenting commuters: thousands of Upshifters who were willing to challenge and change their own behaviors.

But why was this the case? According to Wendy Wood, a leading scholar on the psychology of habits, we don't want to put too much effort into thinking. Things like our route to work become habitual, in that we do them without even thinking about them, responding to cues around us. A bit like Pavlov's dogs salivating at the sound of a bell, the cues of location, times, moods, and people trigger us into automatic, unthinking behaviors. These contextual cues usually serve to maintain the status quo—they capture our attention automatically and lead to habitual responses being triggered. In "normal times," many of us tend not to explore and innovate as much as we can, relying instead on habit and the past to guide our actions, *even* when the problem we are facing is one that is important and matters—when it is "high stakes." In the words of the researchers, commuters "get stuck with suboptimal decisions because of under-experimentation."

But if you change that system of cues—say, with a strike—we are forced to put more energy into thinking about what we do. As Harford concludes, "sometimes the obstacle that has been placed in our path might provoke us to look around, and perhaps to discover that a better route was there all along." To employ an economic metaphor, there were the equivalent of ten-dollar bills lying on the street, which commuters did not notice until the strikes made them apparent.

It turns out that the commuters are not alone in being pushed into a Challenger mindset by stress or pressure. Psychologists examining habits have looked at the impact of major life events on our behaviors and found that big disruptive events can lead to changes in habits. One study examined eight hundred households that had moved recently and that were provided information about new environmental approaches that would reduce their carbon footprint. When compared with settled households, the recently moved households were up to twice as likely to adopt the new approaches.

Another example that the commuter researchers found also related to strikes, this time by police officers in the Netherlands. In August 2015, the strike meant that officers were not available to supervise fans around matches in the Dutch professional soccer league. Some matches were canceled, but others went ahead nevertheless. To the surprise of many, the matches that went ahead were completed peacefully. This taught authorities that a police presence around these events is not always necessary, thereby opening up substantial cost savings. The researchers asked, with mock frustration, why does it take externally imposed events to trigger these beneficial changes? I would argue that Upshifting is at least part of the answer to this. But there is a different question worth asking: why did only 6 percent of people respond to the imposition of constraints and pressure with novelty?

On one level, the reason is obvious: because not all of us are Challengers who can respond to pressure in ways that break with the norms. But what does this actually mean? What do Challengers do that is so different from the average person? Some valuable lessons come from the work of cognitive psychologist Gary Klein, who specializes in what is called "naturalistic decision-making"—not decision-making as we would like to see it, but as it actually is. He spent a decade gathering data about the situations where new ideas are generated, the so-called "light bulb" moments. But he didn't just look at the creative responses. Instead he laboriously went about examining what he called "twinned situations": contexts in which two people—the "twins"—have all the same information and resources, but in which one gains and uses a creative insight and the other doesn't.

Klein illustrates this vividly with reference to a famous incident in the history of US emergency responses: the Mann Gulch fire in the Helena National Forest in Montana, which occurred in August 1949. A group of smoke jumpers parachuted into the burning area, armed with techniques and equipment to stall the flames, originally thinking that it was a "ten o'clock fire"—meaning that it would be under control by ten the next morning. A fatal miscalculation meant that the men were being chased by the fire rather than the other way around.

The foreman, Wagner Dodge, made a radical on-the-spot decision. Instead of trying to outrun the fire raging behind him, he lit his own fire in the dry grass. As his own fire burned a trail up the hill in front of him, it created a safe path in which the larger fire below had no vegetation on which to burn. Dodge then threw himself down into the embers of his fire and crawled up the hill in its wake. Miraculously this kept him protected from the intense blaze all around. His colleagues refused to follow suit and tried to outrun the fire. Two got lucky and found some rocks to cling to. The rest tragically burned alive.

There were sixteen smoke jumpers that day. Dodge's actions put him in the 6 percent who survived—and I don't believe that it is coincidental that this is, give or take a few decimals, the same percentage as the commuters who changed their behaviors as a result of the London Tube strikes. The commuters got to work on time through forced experimentation. Dodge survived by an act of creative desperation. Both he and the commuters were able to bring to the surface, test, and restructure their understanding in a pressured situation. They used the things at their disposal—knowledge and imagination—to change how things were done.

Klein's observations of people in high-stress real-world decision-making situations—firefighters, paramedics, fighter pilots—amount to hundreds of natural experiments in Upshifting, and provide us with some clear pointers as to what Challengers do that is different from those who remain in the zone of cognitive efficiency and social acceptability.

The non-Challenger twins all shared the following: taking a passive rather than an active approach to solving problems; a lack of skepticism or critical questioning; a lack of curiosity or inquisitiveness; taking all information at face value. As a result, these twins were more likely to be caught up in existing, often flawed, beliefs.

The Challenger twins also had a number of commonalities. They "poked around in variability, conflicts, confusions, and setbacks." When things went wrong, the Challengers explored the errors, anomalies, and problems in some depth, and used them as leverage points to seek novel adaptations to the new reality rather than expecting reality to adapt to their beliefs.

They sought "fresh perspectives on problems." Not only would Challengers spend more time speculating about alternative approaches, giving freer rein to curiosity and wondering about promising coincidences; they also sought new perspectives by connecting to people very different from themselves.

And, finally, they embraced and even thrived upon the creative desperation that arises from the urgency of pressure and stress. Indeed, many Challengers seek to stimulate a sense of urgency by thinking through their options in a variety of different scenarios, asking themselves what they would consider if the situation demanded immediate action.

I think one final message to reiterate is that none of us are locked into the 6 percent or indeed the 94 percent alluded to here. People who were once conformists can learn to become Challengers. So too can groups, teams, and organizations.

And sometimes these moments of forced experimentation can guide us into what has been called a "riptide" of change: a cascade of new approaches, ways of thinking, identities, even ways of life. I have even experienced such a riptide myself: in 2013, weighing almost 275 pounds, desperately unfit and addicted to junk food, a three-year-old Koby started calling me "Fat Daddy." The moment was enough to make me sign up for a weekend boot camp, radically change my eating and exercise habits, and lose almost a third of my weight in three years. (Genetics, age, and tummy fat being what they are, Koby now teasingly calls me "Thin-Fat-Thin Daddy.")

One of the most remarkable examples I have found has transformed not just the industry in which it took place, but also our view of ourselves as consumers, and even how we live our lives. It began in 1955, when Gillis Lundgren, the fifth employee of a small Swedish mail-order company, faced a problem.

"WHY NOT TAKE OFF THE LEGS?"

Lundgren had many problems, actually. His company, a recent entrant in an old and established industry, which had only recently expanded its range of products from pens and pantyhose, had run aground despite early successes—in fact, *because* of those early successes. Accused of undercutting prices by the representative trade body, the company found itself on the receiving end of a boycott, with few companies willing to supply to or deliver for them. Their focus on providing goods for poorer agricultural workers who could seldom afford decent items, while laudable, did not win them any plaudits.

That was the big problem. The smaller problem was mundane by comparison: one of the companies that would still supply them had sent Lundgren one of their latest models of tables, which he had to get to a photo shoot for the company catalog. And however hard he tried, he couldn't fit it into his car. Then he had a click moment: "Why not take off the legs?"

This seemingly insignificant decision would transform the fate of that company, and with it the Swedish furniture industry—and, eventually, the very idea of modern living. The company is, of course, IKEA, today the biggest furniture company in the world. Lundgren's moment of inspiration stands in business history like a butterfly's wing at the crest of a globe-spanning weather front.

But Lundgren wasn't the Challenger who set IKEA on the pathway to changing the world. Instead, it was the founder and CEO of IKEA, Ingvar Kamprad, who gave his initials to the first half of the company name. It was Kamprad's vision that led IKEA to move into furniture, and his realization that even in the face of a boycott, IKEA could survive by lowering costs relative to its competitors—and that there were real cost savings to be made in assembly and delivery. (Lundgren, incidentally, was a Crafter, the Upshift archetype we will learn about in the next chapter).

His Challenger insight was that Lundgren had inadvertently crafted a solution to their problem: a way of passing on the cost of assembly and delivery to customers, thereby driving down costs radically. Within a few short years, customers would be doing half of the production of furniture themselves.

To this day, people still marvel at how cheaply IKEA provides furniture and the answer is simple: we do their work for them.

While this worked for a while, soon more problems reared their heads. The boycott lasted, and IKEA had to expand its supplier base to keep up with demand. The company had to look elsewhere for stock. Communist-era Poland was not the ideal business environment, but the manufacturing costs of furniture were 50 percent lower than in Sweden. So, IKEA headed there, taking the legs out from beneath the Swedish— and then the global—furniture industry in the process.

The IKEA case exemplifies the paradoxical nature of Challengers' effect on the world around them. Indeed, one of the most influential business theories of the last quarter century has been built on these qualities. Harvard professor Clayton Christensen argued in his 1997 book *The Innovator's Dilemma* that the cause of many corporate failures was in fact "good management." "Precisely *because* these firms listened to their customers, invested aggressively in new technologies that would provide their customers more and better products of the sort they wanted, and because they carefully studied market trends and systematically allocated investment capital to innovations that promised the best returns, they lost their positions of leadership."

Christensen found that it is thanks to Challengers like Ingvar Kamprad that businesses innovate, flourish, and grow over the long term: "There are times at which it is right *not* to listen to customers, right to invest in developing lower-performance products that promise *lower* margins, and right to aggressively pursue small, rather than substantial, markets."

What businesss Challengers do is bring disruptive technologies to the market, by dint of personal and professional gumption. In doing so, they rewrite the rules of particular products and the customer needs they fulfill. Although their innovations may be less effective in existing mainstream markets (like finished furniture products), they will invariably have qualities that cetain niche or new customers value: they are often simpler, smaller, cheaper, and easier (like IKEA's flat-pack furniture). There are many many other examples: the emergence of small mopeds over large high-power motorbikes in the 1960s, the rise of personal computers over mainframes in the 1970s, the emergence of low-cost airlines undercutting bigger established companies in the 1990s, and how the popularity of digital cell-phone cameras made the camera and film reel industries all but obsolete by the 2010s.

And yet despite the value of all of these innovations, those who suggested them were often disparaged and dismissed. As Walter Bagehot, an early influential editor of *The Economist*, pithily put it: "One of the greatest pains to human nature is the pain of a new idea . . . It makes you think that, after all, your favourite notions may be wrong, your firmest beliefs ill-founded . . . Naturally, therefore, common men hate a new idea, and are disposed more or less to ill-treat the original man who brings it."

If the business world tells us anything, it is that Challengers are vital to Upshifting but they also have one of hardest roles to play in performing under pressure. And nowhere is this more apparent than when human lives are at stake.

"DON'T HIDE THE SCREW-UPS FROM THE WORLD—AND DON'T HIDE FROM THE SCREW-UPS YOURSELF"

Some of us will remember when BBC reporter Michael Buerk gave his broadcast on the Ethiopian famine in the mid-1980s. There was not enough food, not enough treatments for the starving, and people were suffering immensely as a result. There have been few situations— perhaps only the 2004 Indian Ocean tsunami—that have evoked pity on such a global scale. Watching the footage now, with experience of working in humanitarian crises—using hanging scales to weigh children, as if they were goods in an old-fashioned grocer's; the marking of people's heads with an X in felt-tip pen as they sat passively waiting for food—I am struck by how much they were objectified by those responding to their suffering.

One of the most compelling interviews Buerk conducted was with a young Doctors Without Borders (MSF) nurse, Claire Bertschinger, who was responsible for running therapeutic feeding stations where severely malnourished children were treated in the desperate hope that they could be brought back from the brink. Bertschinger had the unenviable task of deciding which children could be treated, and which children were too sick to be saved. She told Buerk at the time, "It breaks my heart." Her account in particular would inspire many around the world to do something to meet the needs of the impoverished and the suffering. Bob Geldof would later remark of Bertschinger, "She had become god-like, which is unbearable to put upon anyone at any point."

Bertschinger would make a rather different association in her autobiography: "I felt like a Nazi sending people to the death camps. Why was I in this situation? Why was it possible that in this time of plenty that some people have food, and some do not?" These were not questions many asked on the front line of famine response in 1984—the sheer magnitude of needs overwhelmed such questions. But they lurked, echoing the famous quote from Brazilian theologian Hélder Câmara: "When I give food to the poor, they call me a saint. But when I ask why they are poor, they call me a communist."

These were exactly the questions Steve Collins was asking himself after his experience in Liberia. The facilities that Bertschinger had been running for starving children in Ethiopia used exactly the same principles and approaches as the ones Collins had deployed for adults in Somalia and were used around the world. With his experiences of Sudan—and keen awareness of the need to engage with and respect communities, no matter how impoverished—Collins decided that the answer would come from "engaging with people better, looking at their strengths, rather than trying to impose solutions on them."

He wrote an article about his failures in Liberia, what happened, and what he should have done. He encountered fierce resistance to the article's publication from the organizations involved. No aid agency wanted to be seen to be causing cholera. But it also meant that lessons were not being learned by the system. Collins had to take the learning on his own shoulders.

It is interesting to reflect on the Challenger approach from this perspective. We are often told that disruptors are uncaring and insensitive, which is why they are able to overcome or ignore the criticism that is inevitably leveled at them. But in the thirteen years I have known Collins, he has always come across as profoundly human and humane. Like all charismatic leaders, he no doubt presents certain challenges to those who work closest to him. But his response to how he kept himself sane and safe in the face of criticism is instructive: he talked about his mum and dad, about the love they had for him, and the security and confidence this gave him to know when he was doing the right thing.

But changing global nutrition practice needed more than just sharp questions and honest reflections. There were two things that needed to change: the practice and the technology. In 1999, now back working in the field, Collins was asked to evaluate a large-scale therapeutic program for MSF. They had a well-known blue handbook on treating malnutrition in field hospitals—Collins remembers it as a cookbook that provided the recipe for setting up and running a therapeutic feeding hospital. Collins found that lack of treatment coverage for the population had caused riots. Each feeding center could treat only one hundred people over four weeks, and there were hundreds of thousands of malnourished children. It was in this evaluation that Collins came up with and proposed a community-based approach: instead of trying to save a small number of people with intensive care, treat many more people through an outpatient approach. MSF shelved the report.

In part, this was justified: Collins's argument was not yet complete. A vital piece of the puzzle was missing. The treatment of adults and children in feeding hospitals required a fortified milk–based treatment, generally made up of milk powder, oil, sugar, minerals, and vitamins. Some were rounded off with sorghum or other cereal flours and needed cooking. This kind of formulation needed both water and refrigeration, which created risks around contamination—as Collins tragically discovered in Liberia. This meant that it needed to be prepared by experienced medical staff. The use of milk also limited the longevity of the treatment, reinforcing the need for feeding hospitals with electricity, refrigeration, and other facilities—not least because the typical recovery time from malnutrition was four weeks.

It was all very well to argue for a better way of treating acute malnutrition than feeding centers, but how to get around the need for the medicines that they administered? Just as IKEA's Ingvar Kamprad needed a Crafter to come up with the idea of flat-pack furniture, Collins needed his own Crafters. They came in the form of André Briend, a development physician specializing in nutrition, and Michel Lescanne, the founder of French company Nutriset. Their careful exploration led them to come up with and test a number of products—pancakes, doughnuts, biscuits—but nothing worked. Then, when laying out his children's breakfast one morning, Briend was inspired by a jar of nutty chocolate spread that had a similar composition of proteins, energy, and lipids as the diet recommended by the World Health Organization

(WHO). He came up with the idea of replacing part of the skim milk powder in the existing recipe with peanut butter. A few short weeks later, the idea was a viable product.

Known today as Plumpy'Nut, it is a remarkable product for its relative advantages over the therapeutic milk that came before it. It requires no water, preparation, cooking, or refrigeration. It has a two-year shelf life, and comes in an easy-to-open foil packet. The production process is simple and cheap, and can be made using crops and technologies that are readily available in the developing countries where malnutrition is most rife. When Collins discovered Plumpy'Nut, the penny dropped.

Armed with the knowledge of Plumpy'Nut, and after a series of small-scale trials, Collins published an article in medical journal *The Lancet* arguing for a community-based alternative to the therapeutic feeding hospital approach. He showed that the hospitals were hugely expensive, resource intensive, and unsustainable. He also argued that the centralization of care undermined local health infrastructure, disempowered communities, and promoted informal congregations of people, risking disease, infection, and death. He argued for an alternate approach, with Plumpy'Nut and products like it at the core of a community-driven outpatient model of treating malnutrition early and fast: kill the monster before it can become a terror.

His challenge was clearly viewed as a step too far. Collins was vociferously criticized by many experts and aid organizations. No one called him a communist, as far as I know. But many said that he was experimenting on children—or worse, killing babies. The accusation was baseless to say the least. Because treating malnutrition at home meant going out and identifying all malnourished children, rather than just those who were attending a feeding center, there was an argument that the approach itself was actually leading to more deaths. This was a risk, as Collins acknowledged. But the fact of the matter was that while there may have been more deaths statistically, this was actually because the method revealed the true extent of the deaths that were occurring. And more important, many more lives would be saved.

Despite the vitriol, the arguments remained largely academic for a number of years, apart from some very small-scale trials. Most of the institutions involved simply had too much to lose by listening to Collins. Here he takes up the story:

> In 2000, I went to Ethiopia. Their government had forbidden aid agencies from setting up therapeutic feeding centres, because they'd had them for years and realized they didn't address the underlying problems. That gave us an ethical opportunity to set up community-based treatment programmes. By the end of 2001 . . . we had treated about 300 cases by giving people ready-to-use food to be administered in the community. The mortality rate was just 4%, five times better than the normal mortality rates.

It turned out that the community-based approach didn't lead to more deaths, despite the much greater coverage. Using the same resources, therapeutic hospitals might have treated fewer than one hundred children, with around a 20 to 30 percent mortality rate—that would have been twenty to thirty children who tragically succumbed to malnutrition. The community-based approach reached twice that number with mortality rates of 4 percent—eight children out of two hundred. In absolute numbers, Collins had managed to multiply the numbers treated two-fold, with far fewer lives lost. The arguments against Collins then perhaps predictably shifted—critics said this was a result of the small sample size, the data was suspect, and Collins and his team were "cheating" by changing the rules early before the children became really ill. Undeterred, Collins and his team carried on collecting data and eventually had information on almost twenty-three thousand cases—and the mortality rate was still under 5 percent.

Others then took up the cause. The largest use of the approach up until 2006 was in Niger by MSF, the same medical humanitarian agency whose therapeutic feeding program Collins had evaluated a few years earlier. MSF medics successfully treated some sixty-three thousand children with a success rate of 90 percent, compared to a 30 percent success rate in therapeutic hospitals. The international president of MSF was reported as saying that it would "do for the worst malnutrition what prophylactics did for malaria." He marveled at the approach: "After 20 years working with medical emergencies, it is a shock for me to see what is possible . . . Acute malnutrition is facing a revolution in terms of treatment." A year later, in 2007, WHO approved Collins's approach and ready-to-use therapeutic foods as the recommended means of treating malnutrition globally.

I want to give the last words on the matter to Collins himself, when I asked him what advice he would give to other Upshifters operating in the Challenger mode. As he put it to me, sunshine streaming in through his study window:

> Try as hard as you can; things don't just pop up perfectly formed. So, you have to go back and change it and have to work out what is going wrong . . . And most importantly you have to own your failures. You have to ask, "Why did I screw up?" and, "How can I improve in the future?" if you are going to have any chance of success. Don't hide the screw-ups from the world—and don't hide from the screw-ups yourself.

This statement goes to the heart of what I have learned about Challengers in writing this book. And it is something that answers questions that we all pose from time to time: What is it that makes some of us question and challenge existing practices? Is it curiosity, wonder, faith, or fun? Is it fear, necessity, or desperation? What is it that gives us the space, the freedom, and the confidence to challenge the written or unwritten rules of how things are done?

At the heart of Challengers is a sense of calling: an intrinsic motivation. If you are driven primarily by external rewards, you are much less likely to be a Challenger, all other things being equal. Being a Challenger means running up against the establishment, and not worrying about your self-image or reputation. Psychologically, it has been found that a strong desire to avoid "ego threatening" scenarios restricts our willingness to search and explore alternative solutions to the mainstream. What this comes down to is our level of emotional dependency on and involvement with others. Remarkably, the Yerkes-Dodson curve can also be found here, describing how our creative performance varies with our level of emotional involvement with others, diminishing at the extremes and peaking in the middle.

What this tells us is that Challengers are not merely people who simply don't care about social stigma. The best Challengers are able to actively manage their fears and insecurities in pursuit of what they believe to be right.

We are always hearing about Challenger types in business and politics, and there is almost a grudging respect for their disagreeable, uncompromising, and often arrogant nature. But Steve Collins's account gives us a valuable insight into what can also lie at the heart of a Challenger: it is possible to be courageous *and* humane, and the key seems to be to make sure we keep asking difficult questions, not only of the world but also of ourselves—and be confident that we can handle the answers.

"JOY LIES IN THE FIGHT"

Lifting up these individual Challengers who sought to transform the practices and policies of organizations, industries, and established experts can be an inspiration to us all. But the adage that power comes in numbers is as true for Upshifters as any other group of humans—and there are indeed many examples of groups, networks, and movements that have sought to change the rules of the game, not just of a part of society but the whole societal and political order around them. Many groups and movements have sought to adopt Challenger behaviors as a collective. One of the best examples of all is from colonial-era Boston. Although that time in history is incredibly well documented, and many individuals have become rightly famous for their roles as revolutionary Challengers to British rule, there is a story that remains largely untold, in part because we tend to look at history through the wrong end of a telescope, magnifying key elements while allowing context to disappear.

In 1770, the city of Boston in Massachusetts was rife with tensions and unruliness. A city of some sixteen thousand people, it had for several years been the location of small-scale rebellions against the British government's imposition of taxes on the colonies. The Bill of Rights of 1689, which established the British Parliament, its relationship with the monarchy, and the role of elections, also stated that no British subjects could be taxed without the consent of elected representatives in Parliament. When Britain started financing numerous military and trade campaigns by levying taxes on Americans, the outcry was almost instantaneous, and nowhere was it more vociferous than in Boston. The events of the Boston Tea Party in 1773 were foreshadowed by years of street protests and demonstrations. In 1768, the British government sent several thousand troops to the city to maintain order. Not only did the military forces start to impound ships on accusations of piracy and smuggling, they also press-ganged sailors into military service. Unsurprisingly, this only served to heighten resentment among the unusually diverse and dynamic Bostonian population of intellectuals, merchants, sailors, and freed slaves.

On March 5, 1770, these tensions spilled over into violence. A fight between a customs guard and a local escalated quickly, and a jeering, jostling crowd formed around the pair, led by a formerly enslaved black sailor named Crispus Attucks. By the time more soldiers arrived, the crowd numbered some three hundred people. After a few ragged exchanges and threats on each side, the soldiers fired an undisciplined volley into the angry crowd, hitting eleven people, killing Crispus Attucks and several others instantly. It started to look as though the situation was going to get even worse, then it was rapidly settled by the arrival of a large number of soldiers and the governor, who called for calm and announced there would be an inquiry into the killings.

The events of that day led to a "pamphlet war" between the protestors and those loyal to the British. One of the most famous of these publications was an engraving, which showed the mob as peaceful, and the commander of the troops with one arm raised, as if making the order to fire. On the ground was the figure of Crispus Attucks, albeit with his skin whitened. He would go on to be named as the first person to be killed in the American Revolution. The engraver of the picture, a silversmith named Paul Revere, would become famous in his own right.

In his defense of the British soldiers' actions, Boston lawyer (and future vice president and president) John Adams was keen that the soldiers received a fair trial, not least so that the British could not use it as an excuse for raising the stakes. In his judgment, he drew particular attention to Attucks' ethnicity and behavior, and the mix of the crowd: "a motley rabble of saucy boys, negros and molattoes, Irish teagues and outlandish Jack Tarrs."

A number of modern historians now see these motley crews as having a profound impact on the revolutionary spirit of the age. According to Marcus Rediker, a professor of Atlantic history, Adams's description was far from unique: "The single most common description of the mob in revolutionary America was as a 'rabble of boys, sailors, and negroes.'"

It is estimated that of the hundred thousand or so American sailors in action around the Revolutionary era, many were sailors who had escaped the harsh life aboard British Navy ships, through desertion or mutiny, and at least a fifth were formerly enslaved people who had escaped or been freed. What they shared was their experiences of escape from brutal conditions and the tyranny of navy captains and slave owners.

These hard-won freedoms were not things they took lightly. Instead, they found ways of sustaining them:

> When they took over a ship, the first thing they did was to elect their own captain. They practiced democracy. They also divided up the loot, the booty, in an egalitarian way. [They] even created a kind of social security system, to provide for their fellow pirates who were injured in battles.

Challenging the hierarchy of centuries of naval tradition, turning ages-old and dominant power structures on their heads, and instilling proto-democratic principles into how they operated at sea was only the start of the motley crews' achievements. The collective Challenger spirit fused with the mutual support they espoused would prove both influential and infectious.

For starters, the transatlantic slave trade is seen as having been limited by the habit of pirates to board slave ships and free the human cargo—if they were willing to become pirates in turn. As Rediker puts it, "Salt was the seasoning of the antislavery movement."

If briefly stalling the horror and inhumanity of the slave trade was not enough, the motley crews were also seen as central to Revolutionary protests against the British in the decade prior to 1776.

In every single protest where the actions of the group went further than the moderate organizers, a motley crew could be found. And Challenger principles from the sea filtered through to efforts on the land:

The crews organised themselves according to egalitarian, collectivist, revolutionary principles. What had once functioned, as "articles" among seamen and pirates now became a Code of By-Laws . . . for their own regulation and government.

The influence of the motley crews can be observed in the behaviors and ideas of those they inspired. For example, the revolutionary Sons of Liberty meted out punishment to government officials who offended them based on ages-old nautical practice: tarring and feathering. In fact, the very word *strike* comes from sailors' actions in this era: in 1768 a protest against a wage cut led to rebellious sailors going from ship to ship and taking down the sails: what was known as "striking the sails." Samuel Adams (a cousin of the lawyer John who disparaged the crews in court) watched a motley crew battling a navy press gang in Boston, and was struck by their actions as "the People's running together for their mutual Defense." He wrote a pamphlet defending their natural rights that would become a cornerstone of the Declaration of Independence.

Today, we can see the actions and approaches of these motley crews echoed in many different settings, from organizational reforms to social transformation. In the following chapter we will see how a group of self-styled Pirates in NASA provide a fascinating example of how spacefaring crews could take inspiration from their seafaring forebears. The story of how three young engineers revolutionized one of the leading scientific and technical organizations in the world is a remarkable one, not least for the so-called "Pirate Code" that governed their efforts, that enabled them to achieve results, be resilient in the face of opposition, and maintain personal responsibility.

The starting precepts of their group were simple and would surely have been recognized by their forebears: challenge everything; break the rules; see risk-taking as a rule, not as the exception; and ensure personal and collective responsibility. We can also see modern-day echoes of motley crews in the many direct-action groups and organizations that seek to use strategic protests to make their cases. The power of direct action comes from what it symbolically says about the implicit and explicit rules that shape our lives when they challenge the status quo on a level that cannot be ignored or set aside.

Few people exemplify this as well as the father of nonviolent protest, Mahatma Gandhi, and there is perhaps no better example of his Challenger efforts than the famous Salt March in 1930—when he walked across India, gathering sixty thousand followers as he went—to pick salt on a beach. This was based on a deliberate, considered strategy of challenging the rules of the British Empire—for almost fifty years, Indian citizens had been banned from making their own salt, but had to buy at exorbitant prices from their colonial masters. For someone who practiced and advocated nonviolence, it is surely the Challenger mentality that Gandhi was referring to when he talked about his experience of the Salt March: "Joy lies in the fight, in the attempt, in the suffering involved."

And this highlights the message that I want to close this chapter with. When we take on the mantle of Challengers, we do not wield influence simply because we are being disruptive—although this may be a byproduct of our efforts. Truly being a Challenger means trying to cast off the shackles that exist in our own and others' minds, and in society at large. And this is at the heart of the difference we make when we challenge: we show ourselves, and others, what it means to be free.

6

CRAFTERS

STRUGGLE FOR NOVELTY

WHEN HOUSTON *WAS* THE PROBLEM

January 28, 1986, is one of those dates that is etched into the collective memory of millions of people. People from all around the world had tuned in simultaneously to watch a remarkable first: a space flight with a civilian on board. Christa McAuliffe had been selected from among over eleven thousand applicants to be the first teacher in space. Bright, dedicated, emotionally open, and with a lovely turn of phrase, McAuliffe represented the modern USA at its best. Her journey from modest social science teacher to space rocket payload specialist had captivated the nation.

Just seventy-three seconds into the launch, the team at Mission Control at the world-famous Johnson Space Center in Houston watched helplessly as the *Challenger* space shuttle exploded mid-flight. The contrast with *Apollo 13* was dramatic. There wasn't even an opportunity to *try* to save the crew. There were no survivors. The inevitable subsequent investigation highlighted another striking contrast. This time, unlike the famous words of the Apollo 13 mission leader, Houston was not merely informed about the problem, it *was* the problem. The *Challenger* accident was the tragic culmination of a series of unseen errors and overlooked failures at one of the most advanced, technologically savvy organizations that human beings have ever established. That bright, cold Tuesday remains the single worst day in NASA's history.

For two of the young engineers at Mission Control that day, those shocking moments would prove transformative. Linda Perrine was twenty-one at the time, and it was her very first day on the famous deck of Mission Control. One of her enduring memories was of a seemingly bizarre report from Cape Canaveral: "They were no longer tracking one target but several." As more information flooded through, her confusion was replaced by the grim realization that it was an accurate description of the aftermath of the explosion. The "several" targets were post-explosion fragments of *Challenger*.

John Muratore, also in his twenties, was similarly shocked, but not wholly surprised. Joining NASA two years earlier from the US Air Force Space Shuttle program, he was tasked with improving the technology that underpinned Mission Control. But as he observed and learned, he realized that the way NASA people used and interacted with software and hardware was the symptom of a deeper malaise.

Much had changed since the 1960s heyday of US space exploration, when the astronauts of the era "wouldn't have swapped places with the Beatles." At that time, NASA was a behemoth of computing power. Buying 90 percent of US computer chips in the 1960s, it also ran some of the largest and most powerful mainframes in the world. If you were interested in serious computing power, NASA was without doubt *the* place to see it in action.

And yet when he joined NASA just fifteen years after the first moon landing, Muratore could see all too clearly that the Apollo systems were—you *had* to whisper it—woefully inadequate. Linda Perrine noted with dismay when she started that the desktop computer she had at home had more capability than the massive mainframes that supported NASA missions. Ironically, many of the companies developing these smaller, nimbler machines were doing so thanks to the kickstart they got from NASA investment in the 1960s. But NASA had not moved with the times. Their core mainframes were slow and inefficient, and incredibly hard to reprogram. Whenever the engineers needed something changed, they had to "wait in line as if it were a lunchtime rush at a post-office with only one window open." In the memorable account of a visiting *Rolling Stone* journalist, NASA Mission Control in the 1980s resembled nothing so much as a 1960s James Bond villain's lair.

The NASA Mission Control screens showed a constant flow of data on hundreds of variables relating to spacecraft launches and trajectories. But any analysis had to be done by capturing information and crunching the numbers manually. Some of the calculations requested by Mission Control staff during a launch would take hours at best to be completed. In some cases, it took up to a month to get the assessments back. In dastardly villain terms, it was as if Blofeld had ordered his evil plan to be executed and then spent thirty days stroking his white cat and waiting for the paperwork to come back to tell him if anything had actually happened.

Muratore's early calls to change the approach got nowhere fast. The standard riposte went along the lines of: "We put men on the moon, we must be doing something right, and besides, change is risky."

But that mentality was itself proving dangerous. In one pre-*Challenger* incident, both the central and backup mainframes got overloaded and crashed during an actual launch, leaving Mission Control completely blind. But the incident was shrugged off, and Muratore and his colleagues didn't have the confidence or the determination to challenge the status quo.

The *Challenger* incident would prove the tipping point. Computers weren't to blame for the accident, but rather the infamous O-ring, a rubberized seal that had failed in the cold winter temperatures of that January morning. NASA management had received numerous warnings and assessments, but they were all overridden, with fatal consequences. The O-ring became an emblem of the same laissez-faire attitude that Perrine and Muratore were battling.

At the heart of both failures—one public and momentous, the other hidden and gradual—was the same ailment that Muratore had diagnosed two years earlier: a culture of "if it ain't broke, don't fix it." Perrine and Muratore knew that they could develop systems using their home computers that would work better, and would be quicker and more user-friendly. Initially hoping that the aftermath of *Challenger* would be enough to provoke new and better approaches, they were bitterly disappointed with the changes that transpired: greater administration, more reporting, more oversight, and more top-down control. And zero changes in management culture or mindset. Those who tried to point out the inadequacy of the response to the tragedy were shown the door.

Enough was enough. Muratore, Perrine, and a few others formed a covert group dedicated to putting their ideas to the test. Calling themselves the Pirates, they started to build their long-imagined alternative to the inadequate mainframe computers.

They began by using their own personal computers but quickly moved onto any equipment that they could borrow or co-opt, and began writing code for an alternate Mission Control system. Some initial money came from an internal innovation fund, which gave them access to the Mission Control data streams. But getting hardware was a continual challenge. The Pirates' way of navigating this was to tap the large number of NASA contractors who were keen to see their machines used at Mission Control and willing to donate hardware to the cause. But because of US government legal restrictions on gifts—read "bribes"—all such donations were on a "test-to-buy" basis. And because the Pirates had no purchasing budget, they had to return all equipment within ninety days.

On the surface of it, this meant a lot of headaches and stress for the Pirates: They had to continually rewrite their software code to be operational on many different kinds of machines. Ninety days was not long to have to deal with equipment switches, especially when you had to do all the work in your spare time, during breaks or late nights around your day job. But the inconvenience actually had a huge upside: Muratore, Perrine, and the other Pirates had to get to know their code inside out and back to front, so that they could efficiently and effectively transfer it across machines of wildly varying technical specifications and keep it working. The ninety-day rule also meant they had to get to know every desktop machine on the market. As Muratore later put it:

> One computer would come in, we'd learn how to program on one computer, and then it would ship out, and then somebody else would loan us a computer. It went like that for years, for two solid years. The advantage being that the code got to be very flexible because we were constantly having to redevelop it.

Another positive constraint was the aftermath of the *Challenger* accident. Although there was no tangible support for what the Pirates were doing, the pause on crewed space flights meant that there was a two-year hiatus when the alternate control tools could be developed, insulated from the high-intensity demands of missions. Slowly, over those two years, the Pirates' motley project took shape, a coalescing and ever-evolving knot of people, hardware, and code, applying trial-and-error to craft a low-cost alternative to the world's most famous Mission Control system in almost complete secrecy.

As their work grew, so too did the number of people involved. Some of them were peers who shared code and data. Sometimes it was just words of encouragement and support. Certainly the most significant of these was legendary NASA leader Gene Kranz, whose role on the Apollo 13 rescue we learned about in chapter 4. Kranz told the Pirates that he had their backs.

Their work continued until one day in April 1988 when they finally brought their "system"—for want of a better word—into Mission Control and set it up on an unused bench. But the middle managers who ran the show wouldn't let the Pirates plug into the mains. After all that hard work, and all that creative problem-solving, it looked like they would be stymied by pettiness and power cables. The standoff was ended by Kranz himself, who mildly requested the middle managers "give the kids a chance." The chance was given, and the Pirates were finally afloat, albeit on a self-made raft.

On that day when the Pirates took their makeshift system into the control center, all the chairs on the established area of the deck were pointing toward the old monochrome screens as usual. All of the operations during this time were simulations and test runs rather than actual flights, so the sense of urgency was less heightened than in the pre-*Challenger* era. But the occasional glances backward and chair swivels by Mission Control staff demonstrated their growing curiosity in the "tinkertoy dingus."

The visual difference between the mainframe mothership and the Pirates' tug was striking enough to warrant their interest:

> on one side of a great cavernous room were the old-fashioned telephone-patch bays, a Medusa of wires curling to the floor, dozens of quaint silver computer-tapes nervously stopping and starting, wooden skids of spent tapes stacked around, miles of paper spewing from immense line-printing machines . . . numbers clicking over on the shabby monochromatic screens cased on steel like old Army-tank parts.

In contrast:

> on the other side of the room were new high-speed disk drives, robotic 85-gigabyte tape-cassette machines and brilliant color monitors screaming with a faint supersonic whine through train leads of numbers . . . [These] seven boxes had replaced an entire building full of tapes.

More significant than how the different systems looked, however, was how the new system literally reoriented Mission Control staff. Curiosity soon gave way to necessity. During a particularly thorny simulation, the mainframe pulled its old trick and froze. The mission director, struggling with his static screen, was suddenly provided a real-time assessment of the situation. When he asked with incredulity where the information was coming from, he was pointed to the Pirates' alternate system, which had not only kept going, but had pinpointed the problem and diagnosed possible rectifying measures. After that, Mission Control chairs weren't swiveling or glancing back at the Pirates' system. They were facing it.

While significant, this still remained largely an on-paper achievement for as long as NASA flights remained grounded. How would the Pirates' system, already getting begrudging interest, prove itself in the reality of actual missions? The opportunity to prove their work would come far sooner than Muratore, Perrine, and the Pirates imagined. The first post-*Challenger* launch, STS-26, was finally scheduled for September 1988. Nerve-wracking for everyone concerned, there was a clear desire to stick to what was known, to run the mission on the traditional mainframes, using the tried-and-tested clunky data systems and approaches.

Then in May 1988, Muratore got a fateful call that would transform not just the fortunes of his group of Pirates, but the whole of NASA, and even space exploration as we know it today.

THE "WORKMANSHIP OF RISK"

Think of the NASA Pirates, building their alternate Mission Control systems in secret in the two years following the *Challenger* disaster, using a hodgepodge of borrowed equipment. Remember Tu Youyou, using ancient manuscripts to try different ways of extracting artemisinin, shifting from boiling wormwood to soaking it in lukewarm water. And Captain Sully, post–bird strike, eliminating all other possibilities before zeroing in on the river and landing his aircraft-turned-glider with the precision of an expert surgeon.

Though quite different on the surface, these three individuals and groups are connected by their unique responses to stress and pressure. They are *Crafters* who successfully develop novel solutions in the face of uncertainty and crises. When Challengers cry, "There must be another way!" Crafters are the people best placed to find that way, and who draw the map for others to follow.

Scholars at the Royal College of Art in London argue that inventive problem-solving of Crafters can be described as the "workmanship of risk." Regardless of the techniques or technologies involved, the success of Crafters is not predetermined, but relies on their creativity, judgment, dexterity, and care. It is through their skillful development of ideas into practical solutions that Crafters are able to achieve the unplanned breakthroughs that are so integral to Upshifting.

Because of their focus on novelty in the face of stress and pressure, Crafters share a great deal of common ground with outstanding performers in sports, arts, and business who undertake bold quests into "that productive, uncomfortable terrain located just beyond our current abilities, where our reach exceeds our grasp."

For our exemplary Crafters, three actions underpin their efforts: deep observation, exploration and discovery, and experimentation. Sometimes this Crafting process takes seconds, as it did for Captain Sully—albeit drawing on years of experience. In other settings, like for

Tu Youyou or the NASA Pirates, Crafting took years. But the actions remain the same and are at the heart of all Upshifting Crafters, whether programmers, pilots, or traditional herbalists.

Either by their own motivation or because of circumstances beyond their control, all of our Crafters found themselves facing a complex, messy problem. Making sense of this invariably means *deep observation*.

Crafters start by "taking stock" of the situation, working to unearth and collect information, knowledge, facts, feelings, opinions, and thoughts. This investigative mindset is core to the Crafters' approach. Exemplary Crafters are fascinated by how things work and why. They're obsessed with the physical, social, and technological phenomena that make up our world.

At the point of stress, pressure, and crisis, Crafters don't simply reformulate threats as challenges. They *fall in love* with the challenges they face, exploring them in depth from a range of perspectives. Think back to Tu Youyou, spending months learning about malaria in affected communities in the tropical areas off mainland China. That deep observation helped her understand more about the enormity of the challenge that Project 523 faced, while building her personal commitment to it. Deep observation was also at the heart of what made the NASA Pirates so acutely aware of the challenges facing Mission Control prior to and after the *Challenger* disaster. Here is Muratore reflecting on the importance of deep observation and learning, and how it can build motivation:

"What's the subject at hand? How does this work? How do we know this works? What's the physics behind that? What's the chemistry behind that? What's the common standard practice that's done in the industry? How do they do it on other programs?" If you're communicating "I want to learn," you can get an awful long way . . . Every day can be a day you walk into a room and you learn from incredibly brilliant people. If you choose for that to be your job. I've always felt that my job was to go in and ask questions and get smart on other people.

One of the most important roles of the Crafter is to redefine and refine understanding of a problem. There is usually some definition of the problem that expresses the "heart" of the situation, and the best Crafters find this in a way that invites novel perspectives. Posing the right question in a new way is often the most important step toward developing novel solutions.

Crafters constantly test the boundaries of what is possible through *exploration and discovery*.

This enables them to generate many new ideas and approaches that may be of relevance to the problem. Not content just with coming up with original ideas, they also assess the resulting solutions systematically.

In the NASA case, once the Pirates were formed, they had to work in constant discovery mode, figuring out how to use the data streams they had access to across a range of different machines, always having to code and recode. Muratore described the discovery mindset that informed the Pirates as "build a little, test a little, fix a little." Instead of trying to figure out the whole problem entirely, they would break it down into its core components, taking them one at a time and not moving on until they had demonstrated that they understood the component fully. Think of Captain Sully using his wealth of knowledge and understanding to run through all of the possible scenarios and excluding them one by one until the river landing was the only one that was left to him and the crew.

Last, Crafters *experiment and prototype* ways of harnessing the phenomena they observe, helping to navigate the crisis being faced. They tweak, tinker, and adapt. The leading global design outfit IDEO—the brains behind the first manufacturable computer mouse for Apple and one of the most successful handheld computers (in the days before smartphones and tablets)—sees the process of discovery as a search across three different areas: feasibility, viability, and desirability. Experimenting and prototyping, in these terms, is the iterative attempt to find novel sweet spots.

Feasibility—What is technically and organizationally feasible?

Viability—What is the longer-term viability (in terms of social, economic, or other goals)?

Desirability—What is desirable?

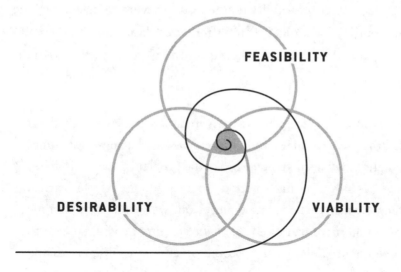

THE WAY OF THE CRAFTER

Experiments and prototypes helped the Pirates to align feasibility ("How can we analyze and interpret mission-critical data with desktop machines?") with viability ("How can we make this work as an alternative to mainframes?") and desirability ("What is it Mission Control *really* needs?"). They enable Crafters to try out novel ideas and processes, going beyond theoretical and in-principle possibilities to make things tangible and real.

Through a combination of creativity and rigor, true Crafters act as their own best enemy. Their work is not merely about remarkable discoveries, although they are often most recognized for these efforts. True Crafters know that the road to joyous epiphanies is paved with judicious edits. There is a joke in my family that I possess exactly half of these qualities myself: I used to be the kind of kid who loved to take things apart—from radiators to radios—but when faced with a mess of spare parts I lost interest, and never got around to putting them back together!

The alternate Mission Control system, the "tinkertoy dingus," was a mess of intersecting prototypes that the Pirates were ceaselessly taking apart and putting back together in different configurations and combinations. Again, Muratore's insight is useful as a lens on the world of the Crafter:

> We built each piece, and we delivered, in a package, something that the users could use right away that would demonstrate functionality they needed, but it broke the problem down so it didn't have to have one big solution for everything. As we learned things, when we added new pieces on, we would take the time to rebuild if we had to, to make the solution general enough to solve the problem . . . If we could break it into little pieces and we found solutions for each of the little pieces—then in four six-month gulps, we were able to bring the whole system online.

These prototypes become the basis of putting novel solutions into direct practical use. Without prototyping, it is impossible to build acceptance of novel solutions. And this is exactly where the NASA Pirates had got to when we left them. They had a viable prototype and some growing acceptance. But their real test was still to come.

We will learn what happened to the Pirates later on in this chapter, but first I want to take a detour via one of my favorite examples to illustrate Crafters in action. It begins over a dinner conversation in Los Angeles in 1942, as America was ramping up its efforts in the Second World War.

THE ARTIST-ENGINEERS

Dr. Wendell Scott, an army medic stationed in San Diego, was visiting two of his old friends in Los Angeles. Tasked with improving medical care in the expanding war effort, over dinner Scott mentioned the challenges he was facing, including one of the most significant. The standard metal leg splints used to rehabilitate and support wounded soldiers were making injuries worse. His friends, a couple who were industrial designers and had recently moved to California, were experimenting with a range of novel ways of using wood.

In his friends' spare room, Scott saw some of these creations and suggested that a wood-based splint could be the solution to the medical problems faced in the care of the wounded. At the same time, it would help to conserve urgently needed steel and metal for other parts of the war effort.

His friends decided to take up the challenge that Scott had posed. They began with a process of deep observation—immersing themselves in the reports and experiences from the front line of battle, speaking directly to wounded soldiers and military medics—to identify the constraints facing both, and to come up with a clear understanding of the problem they faced. The reports from frontline medical officers reinforced the need for an "emergency transport splint" to simplify initial treatment and evacuation of wounded soldiers.

Prior to Scott's visit, they had been experimenting with plywood and molding as a means of developing furniture and sculptures. They had won awards for their work. But they had never produced anything with the immediate functional needs of an emergency splint.

They started experimenting with their materials in an attempt to understand better how wood could be used in new and different ways to support the human body. Using wood for a curved splint meant using wood veneers that could be bonded together with resin glue, then molded using heat and pressure. The curved wooden structures that they were able to produce proved light and strong, surpassing the performance of the usual metal splints. And the required materials were readily available in large, cheap quantities.

The husband-and-wife team spent months prototyping and experimenting with splints of different shapes and wooden combinations—so much so that trying on different glue-covered splints left one of them with hairless legs by the end of the process. After about a year of prototyping and self-experimentation, they had made the first molded wooden splint ever produced.

Flip back to the Crafter visual I shared on page 94. The splint was a new sweet spot between viability, feasibility, and desirability. As a member of the designers' family later put it:

[The splint] was an extremely honest use of materials, wedding [the] understanding of the limits of the material to the functional needs of the splint. Symmetrical holes relieve the stress of the bent plywood, but also give the medic a place to thread bandages and wrappings. The splints represent a perfect example of utilizing to advantage . . . the constraints of a particular design problem. Recognizing and working within these constraints was always key to [the] design process.

Like all true Crafters, our splint designers were characterized by a dogged persistence to see their curiosity through to the end. But they ran into the immediate economic challenge of how to manufacture and mass-produce the splint. Fortunately, thanks to early positive testing and the endorsement of Scott and the navy, they were able to sign agreements with a couple of large manufacturing organizations. They went on to develop the manufacturing process and even the tooling setup for mass-producing the splints. By the end of the Second World War, over 150,000 molded wooden splints had been made and were being used by wounded service personnel.

Buoyed by their material success, at the end of the war the splint designers returned to their first love: furniture. Using techniques that they had perfected with the splints, in particular the shaping of plywood into curves that would align with the shape of the human leg, they developed a wooden chair that was a first: organic, yet molded to the shape of the human body. They also managed to reapply the knowledge of production they gained from military applications toward manufacturing furniture, including some of the same manufacturing contacts. One of the CEOs who supported their work wrote to them after the war saying he thought often of them and their team of "artist-engineers."

The chair that they designed, in whose curved forms the influence of the wooden splint can clearly be seen, is better known as the iconic Lounge Chair Wood. The couple in question are, of course, Charles and Ray Eames, and Lounge Chair Wood was named by *Time* magazine in 1999 as the greatest design of the twentieth century.

Beyond the specific Crafting behaviors, it is also possible to see in the Eameses' work on the splint the ingredients common to all Upshifters. This profoundly influential couple transformed the face of design in the twentieth century with their notions of "creativity in the face of constraints." And they first plied—no pun intended—their trade during a remarkable wartime Upshift.

To my mind, it is this philosophy of the Eameses' work, that shaped so much of twentieth-century Western life, that is so is deeply redolent of Upshifting Crafters.

THE NETS AND SPEARS OF INSPIRATION HUNTERS

Some of us will no doubt be thinking, "Very inspiring, but that's all easier said than done." How does one actually go about trying to come up with novel solutions under pressure? For many of us, even the idea of this can feel a bit like looking at a medieval map: large uncharted areas with the occasional portentous warning that "Here be dragons."

History and experience give us two big messages about the expeditions that Crafters embark upon. The first is that the quest for creative solutions is hard; original ideas are notoriously difficult to find, and those seeking them out need to undertake a wide-ranging, free-wheeling search. The second is that when they come, creative ideas strike Crafters in instances of inspiration, flashes of aha and eureka moments. Crafting is slow, then fast—like a hunt. But instead of just waiting around for inspiration to arrive, the Crafters we have looked at so far in *Upshift* developed the habit of going after it with nets and spears. By examining their efforts, we can get a better handle on what these nets and spears actually are.

There are whole families of methods that exist that enable inventors to come up with new and original ideas in a more systematic fashion. One of the most powerful originated in a patent office in Baku, Azerbaijan, in the 1950s. The Theory of Inventive Problem Solving (TRIZ) was the brainchild of Genrich Altshuller, an engineer who spent decades researching thousands of patents and their underlying technologies. Contrary to accepted wisdom, there are actually common patterns to the journeys that inventors take. Altshuller found that there were around one hundred different ways of solving any given problem and set out to explain each of these. His approach has been used by organizations including NASA, 3M, and Samsung to tackle many different design problems and challenges.

There are many variations, some complex and some simplified, but essentially what advocates of systematic inventive thinking suggest is that all inventive solutions share some common patterns, and we can turn these into tools for thinking about new possibilities. TRIZ has inspired many others to follow suit, often using a simplified and less dense approach.

I have followed this same approach using my own work on Upshifters and produced the following eight-point CRAFTERS checklist. Here are the things that Crafters can do.

CRAFTING ACTION	UPSHIFT EXAMPLES
CROSS-FERTILIZE BRINGING TOGETHER OBJECTS, PROCESSES, OR DISCIPLINES	FLYING A JET AIRLINER LIKE A GLIDER TO LAND SAFELY ON A RIVER
REVERSE INVERTING AN EXISTING WAY OF DOING THINGS	SUPPORTING REFUGEES TO BECOME BUSINESS ENTREPRENEURS INSTEAD OF PASSIVE AID RECIPIENTS
AUGMENT IMPROVING A PRODUCT OR SERVICE BY ADDING OR REMOVING ELEMENTS	CHANGING THE PROPERTIES AND CHARACTERISTICS OF WOOD TO ENABLE ITS USE FOR ENHANCED LEG SPLINTS
FREEWHEEL MAKING THINGS MORE IMPROVISED, DYNAMIC, AND ASSOCIATIVE	PLAYING AN ENTIRE CONCERT ON A BROKEN PIANO, TAKING THE LEGS OFF A TABLE TO FIT IT INTO A CAR TRUNK
TOUGHEN PUSHING TO EXTREME USES TO SEE HOW THE PRODUCT, PROCESS, OR SERVICE SHOULD BE ENHANCED	USING POSITIVE STRESS TECHNIQUES TO IMPROVE EXAM PERFORMANCE; USING COUNTERPHOBIA TO IMPROVE "BIG DAY" PERFORMANCE
EXPAND INCREASING THE NUMBER OF FUNCTIONS AND SERVICES PROVIDED	PERFORMING MORE FUNCTIONS THAN TRADITIONAL MAINFRAMES ON BORROWED DESKTOP COMPUTERS

CRAFTING ACTION	UPSHIFT EXAMPLES
RANGE EMPLOYING A MORE DIVERSE APPROACH TO THE PROBLEM	EARLY HUMAN CAVE DWELLERS ADAPTING THEIR DIETS WITH SEAFOOD AND ROOTS
SYSTEMS TAKING A WIDE-ANGLE LENS ON THE PROBLEM OR ISSUE	FLAT-PACKING GLOBALLY SOURCED FURNITURE AS AN ALTERNATIVE TO A CONSERVATIVE DOMESTIC INDUSTRY; DIRECT-ACTION CAMPAIGNERS CHALLENGING THE RULES OF THE STATUS QUO

Drawn from the efforts of Upshifters embarked on their quest to craft new solutions, these eight points are a toolbox, a way of learning from and applying the lessons of what innovators and designers have done in the past and applying them to a problem we face at the moment.

As a thought experiment, let me give you an example of a problem and show how you can start to apply the CRAFTERS checklist. Imagine you are a pencil designer, and your boss has tasked you to come up with novel products for different customers that will get consumers excited and build their loyalty to your company. Here is how you would apply the CRAFTERS checklist in practice.

- What could you Cross-fertilize? How could you fuse together ideas taken from pens? For example, use liquid graphite pencils that look like pens but allow you to erase what you write or draw.
- What could you Reverse? Pencils make a mark on paper. Could you invert this by selling sheets of graphite with "scratch" pencils that involve removing material from the sheet?
- What could you Augment? How could you augment pencils? What about pencils that combine colors like multicolor pens? How could you build in additional features such as rulers, compasses, and so on?

- How could you Freewheel? How could you use pencils in unconventional ways? For example, could you use graphite pencils as a form of "dry" lubricant for sticky tools such as new keys? Could pencil shavings be recycled as kindling for fires?

- How could you Toughen pencils for extreme or atypical users? How would you need to adapt pencils for use in space, underwater, or in other extreme conditions? How could you learn from this to experiment with the shape of pencils for more typical users? For instance make them easier or more fun to grip.

- What could you Expand in terms of what pencils do? How might you expand pencils to other kinds of tools? For example, would it be possible to make "paint pencils," where the pencil can be used as conventionally expected but can also be dipped in water to create a paint-like effect, but without the usual mess of paintbrushes?

- What could you diversify in terms of the Range of approaches to a pencil? What different materials, colors, styles, and sizes could you think of for pencils, beyond the usual? What would an Alternative Uses Test for a pencil look like?

- What about the bigger Systems perspective? What, for example, are the sustainability implications of the pencil industry? Are there alternatives to the use of wood and graphite? Can pencils be recycled and, if so, how, and with what benefits?

Simply by thinking about new kinds of pencil and using the CRAFT-ERS checklist in a structured way, we can begin to explore the space of possibilities as a Crafter might. We can start to learn that any idea—whether it is designing a pencil or landing a plane—is in essence a bundle of tradeoffs, and we can start to play around with what different tradeoffs might look like. While these are all just back-of-the-envelope ideas, they help point us toward ideas that could really leave a mark, so to speak.

The exact same principles are applicable where the stakes are much higher. Take professional sports performance, for instance. As I explained in chapter 1, innovation and creativity have become central to sports performance at the highest levels. The key ingredients of Upshifting are not only in evidence in elite sportspeople: they are among the most important differentiating factors between the very best and the rest.

In gymnastics, performance is most precisely analyzed and assessed. The International Gymnastics Federation has established a Code of Points, not only to score performances (using the widely known 0 to 10.0 scoring) but also to identify, classify, and assign value to new maneuvers. Every single acrobatic and dance move is listed, illustrated, and assigned a specific difficulty rating and is frequently reevaluated. Many of the skills are named after the gymnasts who are the first to successfully perform them at an official event, such as at World or Olympic Championships. These ratings have become a kind of proxy measure of creative progress in the sport.

Anyone who follows sports will know that the gymnast with the greatest number of skills named after her is US Olympic champion Simone Biles. Biles has upended the sport by doing the seemingly impossible. Indeed, one of her moves, the Biles on balance beam, is so challenging that it has been deemed too dangerous to be repeated, and given a deliberately low rating by the federation to discourage others from trying to emulate it. How did she do this, in the face of a deeply conservative sporting culture and unimaginable pressure?

As Biles sees it, at the heart of her practice is a process of experimental prototyping and adapting: "Building routines, I feel like, is kind of like LEGO bricks. We can take skills out, and put them back in . . . Nothing is right or wrong. It's just you using your imagination and creativity."

Let me show how this works with selected parts of the CRAFTERS checklist mapped to Biles's signature moves and achievements.

- **Cross-fertilization: Balance Beam Dismount** During one competition, Biles achieved the highest score on the beam, despite appearing to lose her balance during her routine. This was, as the judges noted, no mistake: what she had done was to take a tumble maneuver usually found in floor routines and incorporate it into her beam work.

- **Augment: The Amanar** This is one of the toughest vaults for women gymnasts, and Biles was able to push herself some three to four feet higher than the competition, enabling her to complete multiple twists in the air before landing. The result: another high-scoring performance.

- **Toughen: The Biles** Perhaps her best-known move, the first Biles in the Code of Points (there is now a Biles 4 and counting), was developed after a calf injury, where she adapted the traditional move to prevent further injury and pain. It meant finishing the move—involving a backward takeoff and two somersaults—with a half turn instead of the usual full turn, so that she landed facing forward. Many saw this as physically impossible—even after it had first been done by Biles in competitive performance.

- **Range: Vault Yurchenko** The Yurchenko is hard enough—it's a connected series of skills that starts with a roundoff into a back handspring onto a vault. Biles made this the fourth move named after her by radically expanding the maneuver—in the eyes of one observer, "she literally flies." She does a half turn and then twists twice before landing. This vault has the highest difficulty rating of any women's gymnastic maneuver.

- **Systems: Mental health advocate** In a sport notorious for its grueling training and performance demands, one of the most remarkable changes that Biles brought about was to challenge the rigid expectations of the world gymnastics system. After a case of the "twisties"—what gymnasts call the phenomenon of losing control of their bodies mid-maneuver—Biles withdrew from the Tokyo 2021 Olympics citing mental health issues. This kickstarted an invaluable conversation in sport worldwide, especially for female athletes. As she later put it: "The hardest part [was] speaking out on mental health, I knew that I could have the possibility of becoming an advocate for that. But it wasn't my goal. It's not what I really wanted."

Just as with developing her signature routines, a Crafter mentality is evident in how she is navigating this new territory: "I'm still going through my own thing . . . Everybody goes through that process differently and there are different methods that work for each individual person . . . [You have to] use every outlet given to you."

The NASA Pirates would no doubt have agreed.

THE IMPLEMENTERS OF DREAMS

When we left the Pirates in May 1988, they were getting grudging respect and growing attention from Mission Control colleagues. But then their moment of truth would come sooner than they had hoped—or even wanted.

John Muratore had received a call from one of the systems engineers who had identified a major glitch in mission operations. There was a real risk that the rocket motors would fail mid-flight. The problem was that the fault-monitoring tools employed by the old mainframe would not pick up the problem. Instead, it would throw numbers out indicating that all was proceeding according to plan while the shuttle plummeted to the ground.

Muratore and the team had less than four months to build a system that would meet the needs of the flight, and generate the right information, at the right time, in the right format, to inform the decision makers. By comparison, everything the Pirates had done until now had just been toying around. The stakes could not have been higher. A second crash would have been the end of NASA.

They had to work around the clock, but this time, it was mandated and supported by the highest levels of the agency. As one analyst described it, getting the system working in real-world conditions was akin to "a heart transplant patient going under the knife while the operating room is rebuilt." The eventual platform used five consoles, set out very precise fault messages, and even included a built-in rewind function so that any potential glitch could be rerun and analyzed as soon as it happened.

Let's compare the Pirates' work to the CRAFTERS checklist, so we can fully grasp their efforts.

- Cross-fertilize: Bringing in ideas from Mission Control systems in the US Air Force to apply to how space missions were conducted.
- Reverse: Switching the idea that computing had to be done on big machines, putting control back in the hands of engineers.
- Augment: Creating a new visual and analytical layer to provide real-time interpretation of data streams from missions.
- Freewheel: Developing computer code on a range of different systems, making it resilient and able to cope with the demands of missions.
- Toughen: Establishing a system that could keep working and generate solutions even—and specifically—in crisis situations.
- Expand: Creating a single platform to do the necessary analysis and computation and support decision-making.
- Range: Diversifying the kinds of issues and problems that the technology could analyze and assess.
- Systems: Looking at technology not as a stand-alone component, but as part of the wider NASA organizational and managerial culture.

John Muratore's feelings just before the launch of STS-26 in September 1988 are emblematic of the Upshifting click moment.

> My heart was pumping a mile a minute and my stomach was upset. This was the defining moment of the whole effort. Up to that point, no one was going to die if our system didn't work. And this

was the first flight after the accident. We had one shot at it, and then it was going to be the end of everything. We either did it, or the whole organization went under with another accident.

He needn't have worried. During the mission, the Pirates' system was the backbone of the fault monitoring for the first NASA flight in two years. Things worked perfectly, resulting in a successful mission. As Muratore put it, "We had our computer system run in 1988, and it's run on every mission ever since."

Beyond the Crafter behaviors and actions, it is also possible to see in the Pirates' work the ingredients common to all Upshifters. They displayed a kind of creative restlessness, a willingness to question their own expertise, and a dissatisfaction with the status quo. They had the *mentality* that meant they saw challenges as opportunities. Long before they Upshifted to the challenge that informed their most famous work, the Pirates were fascinated with the way things really work. They revelled in the intricacies of the technological and social workings of the world around them. This meant that they were *originally minded* people, best placed to rethink and reconfigure the way things were done. When the opportunity to work on an alternate system arose, their long-established practice of looking under the hood to see how things really work held them in good stead.

Finally, and perhaps most importantly, they were also characterized by a sense of *purpose*. As Muratore put it later:

> Our people aren't our greatest resource. Our sense of mission is our greatest resource . . . When we lose our sense of mission, we are in the most jeopardy. When we have a high sense of mission, we can overcome any obstacle . . . Where we get in trouble is where we lose the sense of mission. We get wrapped up in politics, we get wrapped up in budget and schedule. We get wrapped

up in personal issues. If we want the best for NASA, we've got to keep our mission, focus on what's our mission. If our activity is not clearly, demonstrably, absolutely, most effectively supporting that mission, we've got to change what we're doing. No matter how painful or how difficult. Because our people sense it, and then we no longer get the best out of them.

Their successful mission did not go unnoticed. Soon after that test run, the Pirates were tasked with developing a new desktop-based system for the whole of NASA flight control. Then, when NASA needed to build a new Mission Control Center, transitioning completely away from mainframes, Muratore was put in charge of the entire process. His team adapted a lot of the code and the processes they developed in their first efforts, and the new center went from an empty building to flying the first flight in eighteen months. In 1992, after that second success, the Pirates were asked to lead on the Mission Control capabilities for the largest, most complex engineering project the global community has ever attempted: the International Space Station.

Not only did their system support the numerous initial missions and the process of building the station in space—perhaps the most complex architectural feat humanity has ever completed—but it is possible to see the Pirates' enduring influence in the technology NASA made available to the ISS crew: an array of commercially available laptops that are networked together and adapted to the conditions of weightlessness and the available power.

In 1996, the Pirates received a remarkable honor for what started out as a covert insurgency group: the vice president's Hammer Award for providing outstanding innovations, enhanced performance, and cost savings exceeding $100 million compared to how much it would have cost using the old Apollo-era systems.

They were also profoundly influential: their philosophy and mentality

THE ART OF UPSHIFT

informed the Silicon Valley practice of agile software development. As leading business scholars noted:

> The Pirates' motto of "build a little, test a little, fix a little," regular short-cycle milestones to encourage continuous improvement and experimentation, results orientation, cutting out bureaucracy, encouraging personal accountability and responsibility, and challenging of convention while operating in a large, rule-bound hierarchical organisation were the essence of agility even before the term became fashionable.

The instigators of the ninety-day rule would surely have been stunned to learn that one of its unintended outcomes was to catalyze a creative problem-solving process that would transform the future of one of the most famous of all US government agencies.

What I love most about the NASA Pirates story is how they provide such a powerful example of bridging the gap between reach and grasp that was alluded to earlier.

The poetry fans among you may know that the phrase stems from a Robert Browning poem, in the following lines: "Ah, but a man's reach should exceed his grasp, / Or what's a heaven for?"

I am not sure anyone really knows exactly what a heaven is for, but NASA has surely gotten closer than most. To give the final words to Muratore:

> We have to change the way we do business if we [want] to implement the kind of dreams we have . . . because NASA is in the dream business . . . The thing to do is to keep the challenge on us . . . It's very tough. . . . [but] I think if you keep challenging us, we'll find a way.

7

COMBINERS

REMIX ACROSS BOUNDARIES

THE SEA SLUG AND THE SEA URCHIN

The cold waters off the East Coast of the United States are home to one of the most unusual creatures ever to have graced our planet—and I don't use the word *grace* lightly. It moves like an iridescent aquatic leaf, glowing edges gently rippling as it traverses the currents. It is hard to think of a more enchanting example of nature's beauty.

I have long wondered why aquatic creatures evoke very different emotional responses from their land equivalents. A very dear friend once told me over a delicious birthday meal that prawns are "basically sea cockroaches," putting me off my much-loved pad thai dish for over two years. But few reactions can be more disparate than the wonder that this *Elysia chlorotica* evokes compared with its much-maligned terrestrial counterpart. Because this aesthetic marvel is, in fact, a slug.

The magic that *Elysia* evokes is not limited to its appearance. The vivid green coloring stems from the codium algae on which its remarkable life cycle is reliant. The slug lays its eggs on the green tendrils of the algae, in long sticky strands. When they hatch, the young are black, mottled with red, and they set about ravenously feeding on the algae. Without it they will not become adults. But it is as adults that their relationship with the algae becomes truly extraordinary. The slugs ingest the algal matter through their tongues, which come equipped with tiny teeth. These specially evolved tools are fine and sharp enough to dig into the very cellular structure of the plant, allowing *Elysia* to suck out the green chlorophyll that generates the algal life energy.

Instead of simply digesting the algal cells, the sea slug incorporates them into the lining of its digestive tract, turning its entire body a vivid emerald green in the process. After consuming and assimilating, when its entire digestive tract has turned green with algal cells, *Elysia* floats away from the plant. And this is when the miracle of *Elysia* happens: its mouth seals up, because it is no longer needed. It spends the next year of its life generating its energy through photosynthesis. This last point is so astonishing that it can slip by many people, so let me put it another way: in aquariums, *Elysia* live for up to a year without any food, simply by having lights shone on them for twelve hours a day.

It is not just the chlorophyll that the slug incorporates into its body. While feeding, it also extracts chromosomes from the algae and begins the process of genetically reengineering itself to become perhaps the most unique of all animals. And the algae genes are transferred to *Elysia*'s offspring, lying dormant until they reach adulthood, and the boundary-spanning cycle begins again. As soon as two of the sea slugs mate and one lays eggs, the strange and beautiful magic vanishes, and the adult slugs die, overwhelmed by a virus that experts believe plays an important role in the fusion of plant and animal life. The details of how this happens are not yet fully understood, but the broad biological process is clear: *Elysia spans* the boundary between plant and animal life; it *translates* vital resources and solutions across this boundary, in the form of chlorophyll and the enabling genetic material; and it *remixes* the results into something entirely new: a plant-animal hybrid. Evolutionary biologists believe that *Elysia* is the only multicellular animal ever to photosynthesize.

Absorbing chlorophyll into its gut is not enough to enable it to photosynthesize. Land slugs do this: it's called eating. The biological environment inside animals is very different from that of plants. The sea slug's remarkable feat is not simply transferring and absorbing. It changes the function of the chloroplasts. Too much light can damage leaves over time, and this would also be bad for an animal. The slugs have found a way of converting light to heat that helps to propel them, and is dissipated though the undulations of their frilly wings. There are also by-products of photosynthesis that could be harmful to animals, which plants manage by exuding them through the waterways that are their equivalent of veins. In the absence of these, the slugs have evolved a way of containing them safely. Another compound deactivates poisons that could harm the slug. All of these measures reduce the effectiveness of the algal chlorophyll cells, but make them fit for use in a completely alien environment to the one in which they typically function.

The most astonishing insight that biologists have reached is that what we are observing in the slugs may be the start of a whole new evolutionary branch of life. Early chloroplasts were most likely independent bacteria that were captured and "domesticated" by other cells thanks to their ability to convert energy into sunlight. Similar combinatory processes took place with the nuclei that are the engines of every cell in our bodies and those of every other animal. So, in fact, in performing their combinatorial marvels, the slugs are not acting in ways that are unnatural for life: they are the very essence of life.

One person who may not be completely stunned by the story of *Elysia chlorotica* is scientist and inventor Catherine Mohr. Not only is she an avid diver, but Mohr is also part sea urchin.

UNDERSTANDING THE RULES OF EACH WORLD WELL ENOUGH TO BREAK THEM

Yes, you read that right. In a charming TED animation, Mohr tells of how on a diving trip off the Galapagos, her hand got impaled on a sea urchin, which left a fragment of its spiny body in her fingers. Just before she was due to have it removed, another unfortunate accident while horse riding left her with a broken pelvis. By the time she was able to have the operation to remove the spine from her finger, it wasn't there. Whenever we break a bone, our remarkable body works to scavenge calcium from wherever it can. And in Mohr's case, this included the sea urchin spine embedded in her finger—which had become part of her hip. But Mohr's similarities to *Elysia* are not merely that she is also a species-spanning hybrid.

As one of the world's leading engineer-surgeons, she has built a hybrid professional career *spanning*, *translating*, and *remixing* at the boundaries of human knowledge. Much like *Elysia*, this came about as a series of improbable transformations.

She kicked off her career as an engineer, with the aim of building the fastest cars in the world and racing them in the Australian desert. Years working on alternative energy vehicles and high-altitude aircraft followed, and she got to a turning point that many of us will be aware of: when you get good enough at the technical aspects of a role, you get promoted away from it.

She found herself scanning around for her next challenge. She knew she didn't want to become a corporate leader, which was one of the obvious paths open to her. She had turned down a prestigious offer from the US government, to run the nation's fuel cell innovation program, for exactly this reason. Remembering positive experiences in the area of medical engineering at university, she decided to explore health technologies. A friend at Massachusetts General Hospital arranged for her to observe surgeries and the use of technologies. And the very first operation she observed gave her a clear and life-changing vision of her future path. As with her own operation, it was another procedure that did not go quite as planned.

The patient in question was suffering from an abdominal aortic aneurysm. The aorta is the main blood vessel that leads away from the heart, down through the abdomen to the rest of the body, and is is the largest blood vessel in the body. However, in certain circumstances the wall of the aorta weakens, and can lead to bulges—which, if they burst, cause huge internal bleeding and are often fatal.

Surgeons have developed stent grafts as a way of fixing this. Inserted into the body via an artery in the groin, X-rays of inside of the patient's body are used to guide the graft into place in the aorta. The stent forms a "sleeve" around the bulge, making it less likely to burst.

The procedure that Mohr observed on that day involved a new kind of stent, designed by engineers who were also present in the operating room. The stent didn't work properly and the surgeons had to switch mid-operation to the conventional treatment of cutting through the patient's abdomen to remove the bulging part of the aorta and sew in a synthetic replacement tube. Fortunately the patient survived.

Later on, Mohr sat in on the debrief that followed between the surgeons who had attempted to insert the new stent and the engineers who had developed the stent. From her perspective, the engineers spoke a jargon-laden "geek-speak" that she was comfortable with, while the surgeons spoke in medical-ese, a barely understandable version of English clogged with medical jargon: "It was clear that neither side understood the other."

Her account of the discussion is illuminating, not just for what it tells us about the two groups, but also about Mohr herself. This debrief prompted a life-changing realization: "If you could understand both of these specialized languages and how the human body worked, you'd be able to communicate clearly with both sides and be an invaluable part of the team designing effective devices."

Despite her earlier dalliances with sea creatures, it was Mohr's "life-changing realization" at that moment that set her on the path to becoming a Combiner. That path included graduating from medical school aged thirty-four, helping to develop and advance the world's first robotic surgical assistant, and working to transform surgical care in poor and low-income countries around the world, which is where our paths crossed, as part of a project I set up on innovation in war and disaster surgery.

When Catherine Mohr set out on her voyage into the unknown land of surgery, she didn't travel blind. This wasn't a bolt from the blue or a ricochet change—she kicked off her boundary-spanning work with a research phase to really understand what it would take to be the intermediary between engineers and surgeons. For starters, there was going to medical school; there was better understanding what it would take to do the kind of combinatorial work that she saw as so urgently needed; and she needed to learn from people who were already working to graft ideas from one space onto the other (like everything she says, her analogies are very precise and sharp). Importantly, in these "existence proofs," Mohr wasn't looking to find anyone who had taken the exact same voyage; she just wanted to get a sense of the stages and waypoints.

The learning process was not the only thing that she scrutinized. In taking her first steps toward being a fully fledged Combiner, Mohr also spent a lot of time examining her own motivation and questioning her tenacity for the journey. She did this by setting up a series of tests and obstacles for herself. For example, would she have the motivation to actually begin a medical degree at twenty-nine? She had to do some foundational learning, specifically on the lab and biological side, so she enrolled in night school at UCLA to fill in the gaps. "If you find yourself doing organic chemistry at 10 p.m. at night, that's a pretty good signal of your commitment."

But actually making the shift took the confluence of a few different things. This was not a journey that rested on intellectual pursuits alone. She needed to find alternate sources of financing. In the US, applicants to medical school still have to typically show their parents' tax returns. Here, she tapped her connections, and managed to grab a meeting with the chair of surgery at the world-leading Stanford University, who happened to be putting together a new program on bio-design. There was a potential source of funding: a firm that wanted to do needs assessments for endoscopic surgery, and was willing to put up 50 percent of the fees and a 50 percent stipend for the right student. Mohr aced it.

What this meant was that Mohr was in a unique position from the very first week of starting as a medical student. On Tuesday mornings, when all of the typical students were on core courses, she was able to volunteer to go into the operating theaters, scrub up, and be a researcher-cum-assistant. From her third week of medical school, she started operating every Tuesday morning and she would take her engineering notebook with her every day:

> This gave me a pass to ask the kinds of questions . . . What's the tool? What's the history? Why is it used in this way? . . . People were challenged but people were also fascinated.

Of course, her maturity and professionalism helped. From the outset, she wore her Combiner colors clearly: she went in older; she had worked as an engineer; she was asking questions and trying to solve problems; and she wasn't in awe of the hierarchy:

> I didn't feel that I couldn't speak in an operating room [because] I had some skills and knowledge that the clinicians didn't have . . . I was a learner but a collaborator and colleague in the area of how things get designed and made and modified.

Not only was she exploring the interface of engineering and surgery through her paid work, but it enhanced her core surgical learning. The parallel learning process did not diminish either. In fact, it helped her leap the queue in many ways. But Mohr did have to adapt her learning process to accommodate this. Enabled by the fact that Stanford was very forward-thinking in how it structured classes and took attendance, there was a lot of diversity in Mohr's cohort (as she noted, there were six women over thirty in her year group) and this meant there was a need for flexibility to suit a variety of life and learning styles. Stanford wanted to train the next generation of medical leaders, and they didn't want the system to get in the way. This was perfect for Mohr, who would use videotaped streamed lectures to catch up on core modules in the evening. By the end of her first year, she was in the operating theater three or four times a week, and watching the day's lectures on double speed in the evening.

She started to do some consulting work for Intuitive Surgical, a company founded in 1995 at Stanford. Established with the dream of turning robotic-assisted surgery from the stuff of science fiction into science fact, Intuitive was exploring a number of different technological surgical solutions. Mohr helped the company to rethink the physics of surgeons and robots in procedures such as gastric bypasses for obesity, where she figured out that the arduous and physically punishing procedure for a human surgeon could be transformed through a human using a robot. As she tells it:

> Every single procedure applies the physical and mental capacities of the surgeon to the biology of the human body . . . This sets out the possibilities and the limits . . . You bring a robotic system into the equation, and suddenly you can do a whole set of new things . . . [With gastric bypasses] you can flip the robot around 90 degrees . . . I realized you can more precisely match the kinematics of the robot to the biology of the body.

This was clearly a powerful moment. On the engineering side, Intuitive knew enough about the robot but not enough about the procedures to be able to make the robot work better. On the surgical side, they didn't know enough about the robotic side to know how to do it. At the heart of the combinatorial play was a process of bringing the two worlds together:

> Robot is designed in *this* way, the procedure is designed *that* way, and so we have to ask the question: how can we realign both to maximize the benefits and minimize the shortcomings? How can we understand the rules of each world well enough to break them? What are the rules on the engineering side that are standing in the way of what can be done on the clinical side and come up with better ways of doing things? And vice versa?

Even if only by analogy, one can imagine that this is the same process that underpinned the evolution of the sea slug to incorporate chlorophyll into its system.

Mohr had the good fortune to be working directly with Intuitive's chief medical officer, Myriam Curet, an accomplished surgeon who had both the experience and the gravitas to call for changes in how procedures were being undertaken. They also started to support others to develop their combinatorial skills, helping surgical fellows learn about both traditional laparoscopic bypasses and the new robotic versions, and switching people back and forth to drive the learning curve. The newly designed approach significantly reduced the time under surgery and has since been shown to have fewer complications, all while lowering the physical pressure on surgeons.

This may seem like a slight digression, but it is worth thinking back to the origins of surgeons. The ancient Egyptian and Chinese words for "doctor" were linked closely to surgery—both are said to loosely translate as "remover of arrows." When barber surgeons emerged as a recognized profession across Europe in the middle ages, it was because they had ready access to sharp metal tools and were also well-versed with the necessary coordination skills. As medical and surgical needs have grown more complex and multifaceted, it is perhaps little wonder that engineers—with their access to technology and capacity to undertake precisely controlled activities—are on the new frontier of change.

Intuitive would develop one of the first ever robotic-assisted procedures to be approved by the US Food and Drug Administration, and Mohr would move there after graduating, eventually becoming director of research.

But even as a student, Mohr was doing more than simply allowing surgeons and engineers to talk to one another. She was a Combiner, helping to remake the field.

SPAN, TRANSLATE, REMIX

This is the hallmark of *Combiners* in one vivid analogy: Combiners are people who help ideas, for lack of a better phrase, "have sex." They are innovators not *without*, but *across* borders.

At the core of Combiners' contribution to Upshifting is their ability to see the world as a series of mutually interacting and reinforcing patterns that can be "reproduced and combined." Combiners spot these patterns but they know that harnessing them is not a matter of simply cutting and pasting from one place to another.

The unique value of Combiners when facing pressure and crisis, according to UK innovation scholar John Bessant, my friend and frequent collaborator, is simple: in such moments, they borrow and adapt solutions, rather than start from scratch. Combiners save time and money, open up new possibilities that might not have been thought of before.

Bessant's work on "recombinant innovation" has useful lessons for any of us trying to develop our Combiner skills.

As we have seen with Catherine Mohr and indeed the sea slug, effective combination doesn't just happen, but needs specific skills and capabilities to be developed. In his work in a variety of contexts, Bessant has identified three capacities that directly map onto the Combiner behaviors we have learned about so far in this chapter: spanning, brokering, and remixing.

Spanning and searching—To find analogous solutions in different worlds, we first need to climb a "ladder of abstraction" so that we can see the same problem being solved in different contexts. This is essentially about translating or reframing the problem. Learning to reframe and recombine is a key skill in this process, and one rich with possibilities.

Coincidentally, Bessant gives a great example of turnaround time being a problem for both operating theaters and racing drivers—something Mohr would almost certainly endorse:

> In hospitals it's the challenge of using an expensive resource like an operating theatre as efficiently as possible—swiftly moving out post-operative patients, cleaning, sterilizing, preparing and getting started on the next patient as fast as possible. That has a lot in common with a very different world—fans of Formula 1 will know that a slow pit stop can ruin a driver's chances in the unforgiving world of Grand Prix motor racing.

British Formula 1 (F1) racing driver Lewis Hamilton, who has beaten all records to become the most successful F1 motor racing driver of all time, knows this better than anyone, because Hamilton is not content with just providing insights on the driving side of the equation. Like the best Combiners, he seeks to master both the art and the technology of driving:

Over the years, we've got more and more open minded. You know, engineers often are quite close minded. They are used to doing the same thing they've done in the past in a safe and reliable way because it's worked before. So the last couple of years it's about really pushing, more so on the track, but also pushing the guys into areas where they are not so comfortable. We discovered things that we would never have if we had never done that. It was awesome to break them, open them up to new ideas. That is what has also enabled us to move on to do great things. It's really great to see them not doing the same things as everyone else, thinking out of the box, and it is really inspiring to see them innovate ahead of everyone else.

Brokering—To find ways to move solutions between different contexts, the vital thing is not to make your spanning and searching so high level that the connections are not practically useful. The best Combiners develop the ability to make connections across very different fields, *and* to see a fit between needs and means.

In one of our conversations, Catherine Mohr spoke about how the perception of professional racing drivers has changed:

There was a time when F1 drivers were seen as akin to "meat robots." Innovation was for engineers, people who understood engines and the laws of physics. Drivers were great because they had astonishing reflexes, physical stamina and prowess, and mental acuity, but they were not *technical*. They were like Elvis in rock-and-roll terms: the pure front man with the ability to perform. They weren't being asked to get involved in thinking. But in the modern era, all that has changed. The growth in data, in real-time information, in the team dynamics, means that the setup is more like the Beatles, writing and performing their own music. Drivers matter.

The new generation of drivers like Lewis Hamilton exemplify this change. They have to work at the interface of machine and human capabilities in the face of pressure perhaps more than any other single profession in the world. The very best are fluent brokers between the mechanics of the car, the forefront of driving expertise, and the psychology of competitive sports.

Remixing—Combining is seldom about cutting and pasting from one context to another. Combining means "turning the principles from one world into practices in another," and involves a process of learning and tinkering, or what John Bessant calls "cyclical adaptation."

Here is Hamilton again, in an interview after winning his sixth Formula 1 Championship in 2019. The interviewer has just suggested that his driving technique would not need to change for the next season:

> You can always improve. We didn't win every race last year—there are areas that we stumbled and there are so many elements [in winning]. There's always areas we can improve. As a driver, how I can improve communication with my team and mechanics and the guys back at the factory? But also as a driver, [how can I] deliver better performances? Every year I do. Every year, and even through the season the technique's evolving. This year for example we've got the same tyres as last year, but last year, the whole year was an issue [with the tyres]. So there's still subtle techniques which I'm going to have to adapt, with also a different car. And that's part of the game, is being able to be adaptive, and we've got this new car—I don't know how it's going to handle, hopefully she's great.

Importantly, none of this can happen unless Combiners have "absorptive capacity"—the ability to find, assimilate, and use new resources and knowledge.

With the sea slugs, this is physical, in the form of their digestive tract, which incorporates chlorophyll from the algae. With humans, it is about mental and intellectual capacity, and the enabling environment around us that allows us to capitalize on our combinatorial potential.

And Hamilton, the best in the business, understands this better than anyone. He needs to drive change in the face of pressure. But he can't do it without a cost:

> Diamonds are created from pressure, right? But, I think, for me, I've finally found the right balance . . . I know myself better. I know when to push myself. I know how far to push myself. But I'll back off if I need to breathe.

It is precisely because they spend their time straddling boundaries that Combiners like Lewis Hamilton are so valuable in high-pressure situations. These moments of epiphany can seem random, but in reality Combiners regularly practice boundary hopping between different domains, which increases the likelihood they will find such associations when they are most needed.

Successful Upshifting is frequently the result of Combiners bringing together unexpected bedfellows in a serendipitous synthesis. This echoes the work of Upshifters we have seen throughout this book so far: from how Tu Youyou combined her knowledge of ancient traditional medicine with the pharmaceutical knowledge of counterparts to beat the entire US military medical research operation in finding a treatment for malaria to how French nutritionist André Briend combined aspects of Nutella with malnutrition treatment to make the first batch of Plumpy'Nut.

This set of practices—spanning, brokering, remixing—might sound simple. But the effects can be staggering. Once we start looking, we can see the impact of these processes not just in advanced science, engineering, and medicine, but in the fundamentals of how we live, eat, and drink.

Research undertaken across areas as diverse as language development, gastronomy, and software development found that there is a vital importance of "crossovers" that appear serendipitous while they are happening but actually are the major driver of transformational change in a given sector. And behind each of these crossovers inevitably is a Combiner.

At their most influential, Combiners don't just come up with new ideas or solutions—they create whole new areas of human endeavor. The example I want to turn to next illustrates this perfectly: how a simple and almost-overlooked process developed by a French brewer and confectioner transformed the entire course of human development.

WAR AND PEAS

In 1795 the short-lived French government authority of the time (called the Directory) launched a contest intended to transform the way that the army preserved food for long military campaigns. The prize was a lucrative twelve thousand francs. Among the many professional chefs and food preservation specialists who took up the challenge was a Parisian named Nicolas Appert.

Appert had the kind of varied upbringing and life we might expect from a Combiner. He grew up in his family inn and learned how to brew beer as a child. He opened his own brewery and vintner's operation at age twenty before moving to modern-day Bavaria to become the steward of a count of the Holy Roman Empire. After returning to Paris, he became a successful confectioner, making popular sweets and pastries. He managed to add "revolutionary" to his experience during the overthrow of the monarchy in 1789 and assisted in the execution of Louis XVI in 1793. He was released from his time as a prisoner of the ousted revolutionary forces, who believed him to be too moderate.

After his release from prison, he started to develop an obsessive focus on the preservation of food. Perhaps it was a reaction to the widespread challenges facing the government of the day: one famous quote was that "under Robespierre, blood ran and we had bread; today blood does not run and we don't have any bread." Perhaps it was his own personal experience of prison and the no doubt dire quality of food he would have had to eat for almost a year. (A widespread but almost certainly apocryphal account suggests that he started experimenting in response to a prize announced by the government of the day, but this is one of those innovation myths for which there is scant evidence.) Appert's approach was gleaned not from a single method or based on a single foodstuff. He was convinced that the diversity of his experiences gave him a unique perspective on how a wide range of foods could be preserved. He later wrote of how his experiences of spanning put him in good stead:

> Having spent my time in the pantries, the breweries, the storehouses, and cellars . . . as well as in the shops, manufactories and warehouses of confectioners, distillers and grocers; accustomed to superintend establishments of this kind for forty-five years, I have been able to avail myself, in my process, of a number of advantages, which the greater number of those persons have not possessed.

As a result, he deliberately set out to broker and remix processes from across all of these settings.

The implications of these fledgling combinatorial experiments went went well beyond the culinary world. In the 1790s, Napoleon Bonaparte was leading campaigns across much of Europe. Hunger was the most common experience of Napoleonic foot soldiers. One of the few detailed firsthand accounts of the time indicates little if any concern with the campaigns, the politics, or even particular battles. What

Napoleonic foot soldiers like Jakob Walter really wanted was to avoid horrible injuries and to have enough food to eat. Most soldiers had to buy, forage, borrow, plunder, or otherwise scavenge their own food.

On one campaign, Walter's diary describes how a group of soldiers decide to share their scavenged possessions to make a collective meal, melting a slab of frozen lard and using it to cook up some peas. Only, unfortunately, it wasn't lard, but soap, making the precious peas unpalatable. On the Russian campaign, desperate men would fight over and eat the meat of fallen horses, seasoned with gunpowder.

It will not have been much comfort to Walter and his fellow soldiers on the Napoleonic campaigns to learn that the answer to their troubles was in fact in intensive development as they desperately searched for any means of sustenance. The famous adage, sometimes attributed to Napoleon, that "an army marches on its stomach" may well have been as empty as his soldiers' stomachs. It was Nicolas Appert who would give meaning to these words.

One specific combinatorial insight underpinned his efforts: "If it works for wine, why not foods?" But it turned out not to be quite as straightforward as simply bottling foods, and Appert had to spend some time *brokering* ideas between the two domains. Having studied at length the way in which grapes could be preserved in fermented form in wine bottles, Appert started out by placing a range of different food types inside glass jars, which he reinforced with wire. He then corked the jars, and sealed them with wax. The jars were then wrapped in canvas and boiled, so as to cook the foods.

Each stage of this process needed careful attention. Different foods, Appert discovered, needed to be treated differently. The material nature of the cork he used, and the speed and style of its insertion into the neck of the bottle, mattered a great deal. Appert spent many months learning about the details of corks from winemakers. Some kinds of glass simply cracked under the application of heat, and Appert decided early on that champagne bottles were the best, because of their obvious strength and portability. Appert also found them to be very physically pleasing.

Within a year of starting his experiments, Appert had shut his business and moved to a facility in Massy in the south of Paris. This comprised his home, four workshops, and a yard, where he would grow fruit and vegetables. In his new "laboratory" he tested and trialed a whole variety of foods, including meats, vegetables, fruit, herbs, milk, and whey. Peas appear to have been a particular challenge and he dedicated several pages to their specific challenging qualities, which clearly caused him as much irritation as the soapy version had to Jakob Walter.

Unlike his contemporaries, Appert believed that he could only advance his ideas through "planned methodical experiments, verified . . . by exact observations and [drawing] logical conclusions." This systematic remixing—involving different combinations of foods, materials, and processes in a scientific manner—was essential to the eventual success of his methodology.

Moreover, as Catherine Mohr did much later, Appert saw combinations and recombinations as underpinning the very process that he was testing and perfecting. He was embarked upon a series of quests to "chang[e] the combination of the constituent parts of vegetable and animal productions."

After testing his methods and processes on a variety of foodstuffs, he decided that the underlying principle of preservation was in fact universal: "It operates in the same way and produces the same effects upon all the food material, with no exception."

By the early 1800s, Appert had started to sell his bottled food products in shops. He exhibited them with samples at major food fairs and, most importantly, tested them in collaboration with the French military. The trials of "Appertized" meat, vegetables, fruit, and milk took place across the French navy, with jars being used in ships and in hospitals treating wounded soldiers. Some expressed skepticism, but the overwhelming response was positive—if not gushing. The navy officers' report, which came with a top admiral's endorsement, was that the products had "all the freshness and flavor of recently gathered vegetables." One journalist, sampling his foods, wrote:

> M. Appert was so successful that in each bottle is a bounteous entremets that recalls the month of May in the heart of the winter and often deceives when it is dressed by a skillful cook. It is not an exaggeration to say that small peas in particular, thus prepared, in short are as green, as tender and as savoury as those that are eaten in season. M. Appert has discovered the art of fixing the seasons. With him spring, summer and autumn exist in bottles like delicate plants that are protected by the gardener under a dome of glass against the intemperance of the seasons.

Buoyed by the positive responses, in 1809 Appert applied to the government to claim recompense. The response was not immediate, and when it came, it was in the form of a choice: to patent the process and receive royalties from sales; or to document it, make it public, and claim the twelve thousand francs. Appert chose the latter option: "It is more respectable and adapted to my personality, and more important it is useful for mankind."

Appert's process would have numerous far-reaching consequences. First, and most gallingly for Appert, the same year that he published his public text, it was ripped off wholesale by an English merchant called Peter Durand and used as the application for a royal patent in England, with an expansion of the method to include different kinds of containers, including "cannisters." Produced at large scale in south London by a company called Donkin, Hall and Gamble, commercial operations began in 1813. They gained early support from the military, with the

Duke of Wellington stating publicly "how tasty he had found Donkin's canned beef and recommended it for both the Navy and Army." In 1814, the year before the Battle of Waterloo, the British Admiralty ordered almost three thousand pounds of cans and this soon went up to almost ten thousand pounds.

Donkin, Hall and Gamble would become part of Crosse & Blackwell, still a major food company today. The increased mechanization of can production, together with the introduction of the can opener four decades later (until 1855, cans had to be opened with a hammer and chisel), would prove instrumental in providing sustenance at large scale for people in the rapidly growing cities of the industrial age. Cans of condensed milk would become the first commercially mass-produced objects that were ever purchased in shops. According to historians, this immediately changed the face of new cities, with urban farms disappearing as people switched from fresh to canned milk.

The central role of the tin can in the First World War led to the US government slogan "Back up the cannon with the canner," urging the populace to grow and can food to supply soldiers on the front line. Today, in the US and Western Europe alone, we get through over forty billion cans of food annually—that's approximately forty cans for every single person per year, making Nicolas Appert's method one of the most widespread and replicated processes on the planet.

It was not just his role as a combiner of methods and processes from winemaking to food preservation that was significant. Some describe Appert's "great contribution" as bridging the worlds of chemistry, experimentation, nutrition, and manufacturing as no one had done before: it would be no exaggeration to say that Appert was the first and most influential *food scientist*. Even though he did not fully understand *why* his process worked, he did have a solid understanding of *how* it worked.

Louis Pasteur would take inspiration from Appertization fifty years later to identify the role of microbes in food decomposition and adapted this new knowledge to develop the process that famously bears his name. But it was Appert who had developed sterilization and pasteurization first, as Pasteur himself wrote: "I recognize that this manufacturer made exactly the same trials as mine."

It is little wonder that two hundred years after Appert first submitted his process to the French government to claim his prize, influential food scholars would write that "Appert's work is a stage of first importance in the industrial evolution of our modern society."

The lesson from Nicolas Appert's Upshift stands in contrast to the military leader whose campaigns may have inspired him: the world changes as much because of practical realities as grand ideas.

Which brings me to Albert Einstein's violin.

BELOVED LINA

Albert Einstein didn't just enjoy music the way most of us do. In his later life—when he was one of the most famous people on the planet, albeit for reasons very few people understood—he seldom went anywhere without his battered violin case. It wasn't always the same instrument inside. He owned around ten throughout his life—but he reportedly gave each one in turn the same affectionate nickname: "Lina," short for violin.

Wherever he went in the world, whether to speak with scientific or political royalty, he would take Lina along in the hope of an evening playing with some new associates. At times, even his professional scientific relationships were shaped by music. "Life without playing music is inconceivable for me," he once declared. "I live my daydreams in music. I see my life in terms of music . . . I get most joy in life out of music."

Modern biographers and Einsteinophiles have explored this relationship at some length. Musicologist Nate Campbell, who has studied Einstein's time at Princeton University after his escape from Nazi-occupied

Europe, argues that "[the] making of music offered him a way to go into another mode." Elsa Einstein noted about her husband at work: "Music helps him when he is thinking about his theories. He goes to his study, comes back, strikes a few chords on the piano, jots something down, returns to his study."

Hans-Albert Einstein, reflecting on his father, said, "He had much more the character of an artist than a scientist . . . the highest prize of a theory was not that it was correct but that it was beautiful." Echoing this, a scientific collaborator remembered how Einstein's frequent response to a particular idea was not to critique it but rather to say, "Oh, how ugly." He noted that Einstein was, moreover, "quite convinced that beauty was the most important guiding principle in the search for important results in theoretical physics."

His Nobel Prize biographical entries explain the role music played in this process in more detail: "Einstein's scientific ideas were often firstly created in the shape of images and intuitions, and later converted into mathematics, logic and words. Music helped Einstein in this thought process and helped convert the images to logic."

And the man himself once said that he would never have solved any physics problems without his violin.

In his correspondence with psychologist Max Wertheimer, who was keen to learn about the thinking process behind general relativity, Einstein wrote that he never thought in logical symbols or mathematical equations, but in images, feelings, and musical architectures.

Here he is on Mozart's music: "So beautiful and pure that I see it is as a reflection of the inner beauty of the universe . . . It seemed to have been ever-present in the universe, waiting to be discovered by the master." Another time, he said that both he and Mozart were "musical physicists."

Einstein also echoed his son's perspectives on his own artistic sensibilities: his ideas, he once said, were rooted in his "aesthetic discontent" with existing scientific theories, which "failed to reflect the symmetry, inner unity, and beauty of the cosmos." And then this, in another interview with a psychologist: "The theory of relativity occurred to me by intuition, and music is the driving force behind this intuition. My parents had me study the violin from the time I was six . . . My new discovery is the result of musical perception."

Here is Albert Einstein again, in another letter that is too frequently shortened to a single iconic quote (which I have italicized):

> Words or language, as they are written or spoken, do not seem to play any role in my mechanism of thought. The psychical entities which seem to serve as elements in thought are certain signs and more or less clear images which can be voluntarily reproduced and combined . . . The desire to arrive finally at logically connected concepts is the emotional basis of this rather vague play . . . Taken from a psychological basis, this *combinatory play seems to be the essential feature in productive thought.*

The greatest theoretical physicist of all time clearly saw himself as a Combiner.

Musical physicists. Engineer-surgeons. Algal slugs. Upshifting is replete with examples of combinatory play across the boundaries that exist between ideas, disciplines, and even species.

Being a Combiner doesn't just involve looking at two fields and seeing the similarities, or enabling conversations between disparate actors. Mohr is a great illustration of how Combiners are far more than bridge builders. In one of our discussions, I asked her how it feels to be a Combiner, and to forge a bridge between disciplines. Her answer is typically thoughtful, expansive, and honest:

> The essence of Combiners is that that you can translate between different areas of knowledge—and it is often at the interfaces

where novelty arises . . . You start with common language, data, insights . . . and you can move into that space—the opinion and expertise are set to one side and you actually start to look at the possibilities.

The problem is, though, that most of us are stuck on conventional pathways that are fixed in the middle ground. Becoming a Combiner requires an opportunistic approach that is also wedded to professional courage: not just spanning, brokering, and remixing, but regularly self-interrogating at personal and professional levels, asking, "Do I want to be part of *this* system?" Combiners often have to reject externally imposed measures, metrics, and pathways to success.

Combiners have to work not just at the interfaces of *disciplines*: in Mohr's life and career she has spanned boundaries that exist between technologies, organizations, sectors, gender norms, and above all the expectations that have been placed on her versus those that she has herself generated thanks to her own considerable motivation and passion.

I nudged her again on how it feels:

> It feels . . . well, you have to ensure that your expertise in different domains does not become a trap of learning more about what is *not* possible . . . This means maintaining a "beginner mindset" and steep learning curves in some part of your life . . . As a Combiner, you are very susceptible to imposter syndrome . . . Being a Combiner has to come with humility and recognition about how much you have learned and how much you still have to learn . . . Without that, you won't ever be able to bring the insights from different fields together.

8

CONNECTORS

HARNESS NETWORK INTELLIGENCE

ROS AND MAIA

Ros Ereira is the descendent of successive generations of refugees who fled from persecution in Nazi Germany and Eastern Europe and from the Spanish Inquisition. She traveled extensively as a child and a young woman, visiting many countries around the world, learning languages, forging new connections, and building friendships. Being Jewish instilled in her the firm belief that "you should help a stranger in a strange land." In her twenties and thirties, while working in TV and film, she also volunteered at the refugee drop-in center at her local synagogue in North London. At the critical moment of the European migrant crisis in 2015, Ros would be the catalyst for a remarkable national and international movement to show solidarity with refugees.

When Maia Majumder was a young woman, a book about waterborne diseases changed her life. Maia's family hails from Bangladesh, where, as she puts it, "Annual epidemics of cholera—a waterborne disease—sicken thousands and thousands of people every year." Inspired by the book, and supported and motivated by her family connections, Maia worked at a hospital in rural Bangladesh while still an undergraduate student in engineering, figuring out how to better map and predict the spread of cholera in some of the poorest parts of the country. This would lead her to a career in public health research, focusing on infectious diseases. And at the critical moment of the COVID-19 pandemic in 2020, Maia would convene a unique globe-spanning, multidisciplinary research effort to tackle some of the thorniest scientific questions regarding the virus.

One Jewish, one Muslim, both daughters of immigrants, both keen to remember and honor the experiences of those who came before them—and both catalysts of remarkable collaborations that spanned national, organizational, and intellectual borders. Search for their names online and another similarity will become apparent: their digital footprints indicate two people who have crisscrossed many different social worlds. Majumder is not just a public health specialist; she is also a data scientist, network scientist, science communicator, and race rights advocate. Ereira is not just a campaigner for refugee rights, but a documentary filmmaker, TV producer, historian, writer, and animal health specialist. Both are also professionally exhibited artists, one digital, the other a painter.

In my own interactions with Ros and Maia, it was clearly apparent that networks matter to both of them, not just on a professional level, but also on a personal one. Majumder told me in conversation, "I have no formal training in this space, but I have been an organizer and networker my whole life. I ran for student council, I was the youth group president at my mosque. . . . This is what I am. I do it over and over again in lots of different spaces." Ereira concurs, from her own experiences as a freelancer in many fields, and also highlights another motivation: enjoyment. As she told me, "I *like* connecting people. And their ideas."

When Maia and Ros embarked upon the Upshifts that would become so significant for so many people around the world, their unusually diverse networks, and their ability to activate, mobilize, and transform them at the height of crises, would be at the very heart of the stories. And although their experiences are not unique, as we shall learn over the course of this chapter, they are rather unusual.

CROWDSOURCING SOLIDARITY AND SCIENCE

At just after seven in the evening on September 1, 2015, a message dropped into my social media inbox. An old friend had just sent this message to eleven people, including myself:

> Right. So I've never tried to organise a demo before. But as nobody was doing anything before the EU meeting on refugees, I decided to do something. Which will probably be me and my placard in the rain as I have no idea how to do this. So I'm hoping that some of you lovely people might either:
>
> Have some experience of organising these things and be able to give me some pointers;
>
> Know people who organise these things;
>
> Know people in organisations who might be prepared to have a presence at a demo for refugees (amnesty??);
>
> Have press or social media skills;
>
> Have general words of wisdom??
>
> Time is short and I'm utterly clueless. Please help!! Xxxx

Ever since she could remember, Ros Ereira had been affected by the plight of refugees in the UK. The descendent of successive generations of Jewish refugees from persecution, she spent her spare time volunteering at a drop-in center and a food bank where the clients included refugees and asylum seekers from around the world. She had long been struck by the disconnect between the lives of the people she was trying to support and the dehumanizing language many politicians and media outlets used to describe them, "entirely unconnected to the real people trying to escape terrifying situations in their countries."

Ereira ended up connecting with about a dozen people that evening, and there was an immediate energy. We were people from quite different professions—from theater and journalism to politics and research— and walks of life who shared a common connection to Ros. It became clear that, as well as these shared connections, we were also passionate about supporting refugees.

And we weren't alone: 2015 had seen dramatic influxes of refugees from Syria, Afghanistan, Iraq, and Libya into Europe, bringing in their wake a political and media maelstrom. In April of that year, hundreds of people drowned when a boat capsized en route from Libya to Italy. By the time Ros sent her message in September, over three hundred thousand people had attempted to seek refuge in Europe by making a sea crossing, and almost three thousand of them had died in the attempt.

International organizations like the ones I worked closely with struggled to deal with the reverberations from the humanitarian crises they dealt with in far-off lands arriving on their own doorstep. There is a difference after all between dealing with humanitarian missions at a distance and navigating thorny political terrain at home. But there were some notable exceptions such as Doctors without Borders, UNHCR, and the Red Cross.

Small-scale charities and volunteer movements sprang up to meet this sharp spike in human need and suffering, providing food and shelter and arts and education in refugee camps across Europe, even patrolling coastal waters and attempting to provide safe passage. But for many in the UK and across Europe who, like Ros, saw refugees as fellow humans who were suffering, there was no real outlet to show their support.

Until now. The social media page that Ereira had set up started gathering large numbers of likes and shares. By the end of the day on which she messaged us, almost one thousand people had signed up. And it kept growing. Ros's fear of waving a lonely placard in the rain had all but vanished the day after when she got a call from the Metropolitan Police of London, politely inquiring what exactly she thought she was doing, not consulting them on organizing a march through the UK capital that looked like it would exceed ten thousand people.

The discussion on the group chat continued apace and was wide-ranging: from what the campaign should be named, to its route through London, to the kinds of speakers and performers we should have, to the demands we should make on the march. As the numbers grew, debates were had on how we should organize all of this. Despite the extraordinary growth in the number of people signing up for the march, the discussion continued to be conducted in a remarkably civil, courteous, and respectful manner. Then on Thursday, September 3, a single photograph meant that all hell broke loose.

Fast-forward five years to March 2020, and the entire world was in the grip of a different crisis: not catalyzed by desperate mobility, but by its opposite. The COVID-19 pandemic is still unfolding as I write this, and it is a strange and surreal period in human history. For many people I know, it has the continued feeling of a bad dream from which we are still hoping we will collectively wake up. As we all know, the lockdowns called for in almost every country in the world have affected people very differently, in terms of collective and individual mental health.

For Maia Majumder, then a public health scientist in Boston, the shift in working patterns during the start of COVID-19 lockdowns in March 2020 was not especially disruptive or indeed surprising. She had been working on COVID-19 since January as part of a collaboration between Harvard Medical School and Boston Children's Hospital. It became obvious to her quite early on in the year that COVID-19 was not going to go away, and that its speed and efficiency of transmission, together with lack of necessary medical advances, would mean that it would soon be affecting everyone on the planet. Specializing in computational epidemiology meant that staring at computers for large parts of the day was already normal for Majumder.

But she soon became aware that there were many researchers and scientists around the world and in the US who were not so lucky— either because of the lack of facilities afforded her by Harvard and Boston, because of the impact of sudden digitalization on more "analog" disciplines, or because of the effect of the pandemic on academic life, especially early-stage researchers. As she puts it:

> Laboratories were shutting down. Clinicians could no longer see their patients. The postdoctoral job market had suddenly dried up, and many recent graduates were concerned about the gaps the pandemic would leave in their CVs. Even among those who still had work to do, there was a feeling of listlessness: Everyone wanted to contribute something to the fight against COVID-19, but some worried they didn't have the ability to do so on their own.

In Maia's mind, this amounted to "pent-up brain power." With her background of studying COVID-19's predecessors—SARS (severe acute respiratory syndrome) and MERS (Middle East respiratory syndrome)—Maia knew all too well that the virology of the disease was complex and hard to understand, and that the public health response in the initial stages would need to be as much about social and economic responses as medical ones. Like with SARS and MERS, the questions public health specialists were asking were quickly moving beyond epidemiology to the human and social aspects of the response. And it was exactly these scientists who were most limited by the lockdown: these "benched" academics who had vital ideas and thinking of relevance for the pandemic response—if only they could be brought together.

Here, too, there was precedent. During the SARS epidemic in 2003, web-based collaborative technologies and open collaboration across research laboratories around the world meant that the virus could be detected and analyzed in record time. It was also personal for Maia. She had a life-changing experience studying and working in Bangladesh, which made her realize how important it was to burst the self-referential bubble of elite US academia. She decided she had to try to do something about the situation.

Five days after the US national public health emergency was called, Maia put out the following call on social media:

> If you're a trainee with skills relevant to . . . #COVID19 research, please email me your CV. I met my earliest scientific collaborators on Twitter, so it's time to pay back the opportunity.

In a follow-up message she wrote: "Relevant skills, dedication, honesty, and kindness are the only requirements."

Her vision was that, whatever emerged, it would enable researchers to come together to answer the really tough, complex interdisciplinary questions that COVID-19 was posing, and that it would also provide a platform for researchers who had the desire to contribute and had the skills and resources, but not the necessary networks.

Maia was flooded with responses, from across disciplinary and national borders, including medical doctors, psychologists, veterinarians engineers, lawyers, and AI specialists. As she told me, the scale and speed of the replies "made me see I had hit a nerve." Together with her husband and a friend, she set up an oversight committee for what would become the COVID-19 Dispersed Volunteer Research Network. Like the refugee network Ros had harnessed, Maia's hastily formed network faced a crunch point: she decided to organize an online "hackathon" to bring the embryonic members together. This would be the first test of whether as part of the network equation it could be more than the "net" and get down to the vital "work."

WINNOW OR WIDEN

Are you the kind of person who reaches out or hunkers down when you start to feel the pinch of pressure and stress? Cognitive psychologists have been asking this question for some time now and finding answers through both experiments and a number of large-scale empirical studies.

Three of the most extensive studies looked at stressors that many of us around the world have had to face with growing frequency in recent years: job insecurity, financial shocks, and widespread lockdowns as a result of the 2020 COVID-19 pandemic. Each study investigated how different groups of people harnessed their social networks before, during, and after the shocks and stresses in question: how typical workers responded to job insecurity; how professionals dealt with unexpected price shocks; and how ordinary citizens coped with the lockdowns.

If you are fortunate enough never to have experienced the uncertainty and risk of losing your job, you should know it is quite the multiheaded hydra of emotion, with different anxieties hitting you at different points and in different places: anxiety about money, worries about future career prospects, embarrassment, feelings of social shame and worthlessness. Looking at social and psychological data from people working in a range of industries and at different seniority and salary levels, it is possible to see common patterns in how people respond in terms of how they use their social networks. There are two extreme reactions: one common, and one less so. Most of us winnowed—we accept the situation, and the resulting emotional difficulties make us withdraw from everyone apart from those closest to us. A small proportion, however, widened: we broaden and expand our networks, and actively try to harness more diverse connections in order to find a solution, whether that is through establishing greater job security in our current organization or finding opportunities with new organizations.

We have all had to live through an extraordinary couple of years, and many of us will be only too aware of the impact on our social and personal lives. A recent study looked at ordinary people and how their social networks changed as a result of the COVID-19 lockdowns. The researchers examined the networks of a representative sample of US citizens in 2019 before the pandemic and again in 2020 during the March lockdown. The same pattern was identified as with the job-insecure groups: many people winnowed, and some widened. While all groups saw a shrinking of their networks, there was a pronounced and remarkable gender difference: women's networks on average shrank by one twentieth (5 percent), while men's networks shrank by almost a third (29 percent). While these differences didn't explain job success, like the earlier example, they did have a clear impact on the mental health of the participants.

Let's take a moment to think about these findings in relation to our own lived experiences. Some problems feel like a private prison until sufficient courage, desperation, alcohol—or a combination of all three—lead to a temporary parole. We might be more amenable to socializing other problems, openly sharing our fears and insecurities, as well as exploring the best ways of navigating them. If you're anything like me, the social path you take through stress and pressure would seem to depend on the nature of the problem, and the people immediately around you when you experience it. This would seem to reinforce the research findings.

But there are also some more surprising findings from the research.

First, the studies were extrapolated and the same pattern found to hold true for individuals, groups, and organizations experiencing all kinds of sudden stresses, big and small: from families experiencing bad news to law enforcement personnel dealing with a terrorist attack to disaster responders dealing with natural disasters. Regardless of the group or the challenge faced, we see a clear and consistent pattern of most people turning inward to their closest contacts, but a small number opening up and more successfully navigating the challenges they faced. Why should this pattern hold across so many different settings?

One answer comes from the world's leading light in network science, Hungarian American researcher Albert-László Barabási (incidentally a prominent member of a network science research group of which Maia Majumder is also a member). Network science, as its name suggests, is the study of dynamic networks—these might be social, biological, or physical. In a social network, each "node" is a distinct person, and "links" are the connections between each person.

He argues that networks drive success in many different fields, but especially whenever uncertainty means outcomes and success are unclear and hard to predict. This reinforces the job loss and COVID-19 lockdown research findings that in situations of pressure and stress, networks can be *the* difference that makes *a* difference.

But Barabási goes further: the factor that makes a difference in these networks is the presence of individuals with heightened levels of *network intelligence*. As he puts it:

> Networks brim with opportunity . . . because they're held together by . . . connectors [who] are eager to uilize their relationships to support people and causes they find value in. They're especially good at seeing opportunities in the social fabric that other people miss.

In my own work looking at individual case studies of people responding to extraordinary events, I have identified the same pattern: small numbers of Connectors like Ros Ereira and Maia Majumder who are able to maintain, change, and even grow their networks in the face of crises and disasters.

This pattern begs the question: *why* are Connectors imbued with network intelligence, whereas most of us experience its opposite? And following on the heels of that: Is the network intelligence of Connectors something that can be analyzed, understood, and replicated? Can the winnowers become wideners? Can we learn to be better Connectors? It turns out the answer is yes.

THE SOCIAL LIFE OF UPSHIFTERS

As well as analyzing the range of individual responses to shocks, the "networks under stress" researchers sought to find out what influenced these responses, in particular the positive outliers. People displaying network intelligence under pressure were not influenced by the size of their networks, nor even who was in them. Instead, the critical factor was their *perceptions* of these relationships.

The best indication of whether people winnowed or widened was not the size of the their networks, but how they thought about, remembered, and used their available networks.

During lockdown, the women studied had on average of a thousand people in their available networks, but their networks shrank by around fifty. Men on the other hand had almost 1,500 people, but lost around five hundred of them on average.

The researchers suggested that this was because women seem to have greater ability to recall and maintain networks at times of pressure—whereas the majority of men winnowed.

This is reinforced by research in a very different realm from either those experiencing job insecurity or lockdown isolation. Specifically, Wall Street traders facing sudden changes in market conditions. It has almost become the emblem of financial sector excess: selfish, self-centered executives shouting "Buy, sell, hold!" into mobile phones while screens behind them flood with red digits. The reality, however, is that the best traders have their own version of winnowing and widening in the face of shocks: some reach out to more people to make sense of what is happening, while others panic and bunker down.

Analysis of some nine thousand traders inside and outside a medium-sized US hedge fund—examining in-depth analysis of their trading actions and some twenty-two million messages sent by almost nine thousand people inside and outside the firm—revealed an identical pattern of responses to price shocks as among the people experiencing job threats or social lockdowns. The majority of traders winnowed. But some widened—and the vast majority of successful traders were among the wideners. Remarkably, how they used networks was a far more powerful predictor of their financial success than *any* other measurable indicator—including their jobs, existing financial positions, seniority in the firm and levels of experience.

Their mentality—the positive or negative attitude to their networks, and their ability to approach dramatic changes through analysis, insight, and reflection, as opposed to panic—made a major difference.

Interestingly, the relationship between traders' networks and their mindsets was an example of a virtuous circle: having a positive mindset affected how they harnessed their networks, but as these networks were harnessed by specific traders, their state of mind and positive mentality was enhanced in turn.

Remarkably, recent experimental research has borne this out. When we are under pressure, focusing on the density and extent of our existing social networks—as opposed to imagining them as sparse and disconnected—can enhance our ability to cope with life stressors. Simply by *visualizing* our friendship ties taking a more dense form around us, we can better build our internal resilience and support.

Across all the studies, then, it was the *mentality* of respondents that had the biggest influence on how they capitalized on their networks in moments of pressure and stress. Seeing problems as a challenge instead of a threat led Connectors to envision their potential networks of support and insight. They brought new coalitions into existence through the strength of their social imagination. The most successful literally *connected* their way out of crises.

In other words, the different ways in which people use their social networks during crises is explained by reference to one of the core concepts of Upshifting. When we experience stress and pressure, as we already saw in part I, we have the option to frame it as a challenge or a threat. It turns out that this can influence many aspects of our thinking and behavior, including, remarkably, our social relationships.

But how exactly does this Connector Upshifting work? How do people with network intelligence "see the opportunities in the social fabric"?

Imagine you are in one of these groups and facing a major stress point. Your cognitive state can be represented with the three-step sequence in the diagram below.

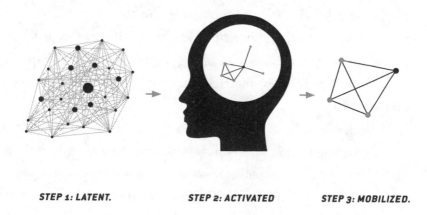

STEP 1: LATENT.   STEP 2: ACTIVATED   STEP 3: MOBILIZED.

Step 1 is the state of your Connector brain before you face a threat or challenge. As you go about your business on a daily basis, in your mind there lies a huge "potential network"—a kind of latent social structure that represents everyone you might *possibly* call upon in a given situation.

Step 2 kicks in when specific scenarios present themselves—an email from senior staff about job cuts, say, chatter about an imminent COVID-19 case spike and the looming threat of a lockdown, or an early indication that a stock's price is about to change. At this point, your potential network crystalizes in your mind into a "cognitive activated network."

Whether you adopt a positive challenge or negative threat stance profoundly shapes the size and structure of the networks that you activate and mobilize.

Step 3 is what happens if and when you reach out to specific connections for ideas, support, and advice. At this point, the people you reach out to become your "mobilized network"—the small kite on the right-hand side.

These three cognitive steps help to shed useful light on how Connectors harness their social networks under stress and pressure.

When I was writing this, my son, Koby, was in one of three stints of homeschooling during the pandemic lockdowns. He took one look at this over my shoulder and said, with a sense of familiarity, "Ah, the friendship fish." His apparent recognition of the image made me think that maybe he had done something about social networks at school. It turned out that he was just describing what he saw as the shape of the network in Step 2. But he inadvertently provided me with a rather useful metaphor.

If we are in a threat state, it can serve to hijack our potential networks. We winnow down to a few small fish, to extend Koby's metaphor. By contrast, in a challenge state, we take a more expansive view, and think about possibilities that we would not otherwise have done.

Connectors don't think in terms of a single friendship fish, as Koby suggested, but rapidly sift through fish of different sizes and shapes and consisting of different ties. Some might even imagine whole *schools* of fish.

In other words, Connectors are people who can Upshift their way to new social configurations and combinations: gifted with network intelligence, they are able to activate and mobilize more network configurations than the rest of us. Our brains often tell us to winnow down, and unfortunately, most of us listen most of the time. This explains why we are not all equal networkers, especially in situations of uncertainty and pressure. But we do also have wideners among us, who are better able to "see the potential in the social fabric" and capitalize upon it.

Interestingly, I was very aware of this when speaking with Maia Majumder about how she saw the COVID-19 research network operating: I could almost see the different fish-shaped networks activating in her mind as she spoke.

BRIDGES, BONDS, AND LINKS ON PAUL REVERE'S MIDNIGHT RIDE

In the early 1990s, historian David Fischer decided to explore a seminal moment in American history: the midnight ride of Paul Revere. In 1775, five years after his famous engraving of the Boston Massacre went viral (as we learned in chapter 5), silversmith Paul Revere set out on a ride. His midnight journey has of course become immortalized in US history and poetry.

Upon assembling a team of graduate researchers and starting to gather initial data on the ride, Fischer was struck by two things. First, there was a huge wealth of primary data that could be assembled to get a more accurate and nuanced picture than had been provided by poetry and myth. And second, despite its significance in US history, there had never been a serious historical investigation of Revere's ride, what happened, and its consequences.

Of course, Revere had long been seen as influential. But it was the nature of his influence, and how he came about it, that fascinated Fischer. The popular retelling of his story dated to the 1860s and was of a solitary hero acting in the interests of the fledgling American nation.

It was told in a way that resonated with the Civil War sentiments of the time, and Fischer also noted that this was the archetypal representation of an American success story: the lone individual who makes a difference in an unfriendly world.

As Fischer would discover, the reality of Revere's ride was rather different.

Piecing together data with his team of students at Brandeis University, Fischer started to uncover a new picture that showed Revere's influence came not from being a solitary agent, but rather from his role as a highly networked organizer. By examining the records, Fischer found that more than sixty people rode that night, and that Revere was the facilitator of "an integrated network of individuals coming together to fight for common goals." Or in other words, a Connector.

On the ride, Revere's actions triggered instantaneous response in the towns he visited, while one of his contemporaries, William Dawes, rode out with little or no effect:

> Along Paul Revere's northern route, the town leaders and company captains instantly triggered the alarm. On the southerly circuit of William Dawes, that did not happen until later. In at least one town it did not happen at all. Dawes did not awaken the town fathers or militia commanders in the towns of Roxbury, Brookline, Watertown, or Waltham.

Fischer's work led to a popular reappraisal of Revere and his contribution. A few years after Fischer's book was published in 1994, Malcolm Gladwell famously used Fischer's account in his influential book

The Tipping Point, where he cited Revere as an archetypal Connector, making the comparison to William Dawes, whose networks were not as expansive or deep.

Other researchers and scholars followed Fischer and Gladwell, including network specialists, and showed that Revere was indeed someone uniquely able to "link up the far-flung revolutionary dots." Revere was a signed-up member of more rebel organizations—including the well-known Sons of Liberty and the St. Andrews Lodge—than any other individual in colonial Boston.

Subsequent investigators have explored the data Fischer uncovered using formal mathematical techniques, revealing a more detailed picture of how exactly Revere harnessed his connections. Reinforcing Fischer's argument and Gladwell's subsequent elaboration, when we look at the network map of some 250 Boston-based revolutionaries operating at the time of the midnight ride, Revere is by far the strongest bridge between the different groups. He is more than twice as important a bridge than the next-ranked individual in the data.

But the data that Fischer and his team uncovered didn't stop there. Bridges between diverse groups are not the only type of social ties that count. It also matters who is closest to the greatest number of people *within* particular groups. This allows a different kind of influence: not based on bridges, but on *bonds*. Here, too, Revere came out top of the list. He didn't simply connect people *across* different groups, but he also knew more people *within* each group.

Last but not least, it also matters who knows the most well-connected people and can therefore make use of their influence. On this factor of *linking ties* Revere also came out joint-top of all of the 250 Boston-based revolutionaries. As a result of Fischer's analysis, we can see Revere was far more than just a solitary actor.

Moving away from an individualistic lens on his achievements to his ability to harness collective action doesn't detract from Revere—but instead highlights exactly how remarkable he was. He connected people through *bridges*, *bonds*, and *links*, using his network intelligence to great and lasting effect.

Revere was "a great joiner . . . an associating man" who used relationships, alliances, and networks intelligently to catalyze a connected movement that would change American history—and with it the world.

The reality of Paul Revere's efforts revealed by modern scientific techniques gives us a clearer picture of how Connectors work to generate collective benefits in the most challenging of times. And as we will see next, his equivalents can be found in virtually every crisis, whenever and wherever they happen in the world.

THE TIES THAT SAVE

Many readers might remember waking up the day after Christmas in 2004 and seeing the horror wreaked overnight by the Indian Ocean tsunami, which struck fourteen countries and killed almost a quarter of a million people. As well as the record-breaking devastation, it still stands as the most well-resourced disaster response of all time, with over $14 billion raised in a few days.

I was involved in helping to coordinate the immediate response, supporting the sharing of learning and innovations across the affected countries, and played a role in the international evaluation that followed, which was chaired by former US president Bill Clinton, and which remains to this day the largest cross-country post-crisis investigation ever undertaken by the international community.

This work and the lessons from it underpin a question I have posed in numerous talks and events over the years: What is the single most important resource in a crisis or disaster? Medicine, some say. Food. Blankets. Cash. Time. Some more complex responses might be functioning markets, effective government, or a free press. All good answers, but all wrong. The correct answer is in fact social bridges, bonds, and links.

In his foreword to the Tsunami Evaluation Coalition report, President Clinton wrote:

> Local structures are already in place and more often than not the "first responders" to a crisis. The way the international community goes about providing relief and recovery assistance must actively strengthen, not undermine, these local actors.

Disaster scholars increasingly refer to the social ties and structures that people use for physical, financial, and social support in times of crisis as "social capital." This is based on the idea of better understanding the different *types* of connections that exist between people, and gives us insight into the particular qualities and benefits that these ties can bring to individuals in groups, organizations, and communities.

The idea of social capital was first popularized by Robert Putnam, the Malkin Research Professor of Public Policy at Harvard University, who described social capital as "connections among individuals—social networks and the norms of reciprocity and trustworthiness that arise from them." Scholars of social capital have identfied three types of ties—and perhaps unsurprisingly they are exactly the ones that explain Paul Revere's influence in Revolutionary Massachusetts.

Bridging ties connect members of different social groups, like unions, nonprofits, and volunteer organizations, facilitating civic engagement, reducing conflict and violence, and providing mutual support across those groups. You will definitely know the bridgers in your own community—in fact, everyone around you will know them, because that's how they work. The purported six degrees of separation is meant to be an *average* count of the number of social linkages between any two people in the world. Dig into the math, and almost any network—whether you're talking about the number of Hollywood actors connected through Kevin Bacon or the number of Facebook users who are connected through Brooklynites—relies on a small number of people whose greatest strength is to be bridgers: those who hold seemingly random ties that span different cliques and groups.

It is widely acknowledged that many challenges can be better understood and navigated when the key players are from a range of different experiences and backgrounds. Traditional hierarchies are generally a less-than-ideal way of working under pressure and stress because of their emphasis on predictability and repeatability: they can often fail to change until they are forced to, by which time it is too late. Being unafraid to let outsiders in is an important way of shaking things up—and it can be essential during crises.

Bonding ties connect members of the same social groups, like family members, neighbors, and members of co-ethnic or co-religious groups. The strength of these ties help those groups survive crises. When faced with ambiguous, complex, and uncertain challenges, and where goals may be fluid and not fixed, we can't just rely on diversity: we also need to reach out to the people we can trust and rely on.

Finally, *linking ties* connect ordinary people to those with power and influence—for example, to local, regional, and national officials, helping them access key public goods they might not otherwise receive. These connections of influence and patronage have a profound influence on how people cope and thrive during crises.

When an emergency happens—whether it's a sudden-onset disaster like a hurricane or an earthquake that can strike in minutes or hours, or a slow-burn crisis like a conflict or famine that builds up over months or years—social connections have the biggest impact on how well people cope and survive.

People rely on one another in crises—on their relatives, neighbors, friends, and communities. Social connections have proven essential in every humanitarian crisis I have been involved in, from Sudan to Syria, and from New Orleans to New York. Time and again, it has been found that even the most vulnerable communities with little access to material resources or support can respond positively to crisis if they have strong social networks to rely on.

The most startling finding in our cross-country evaluation of the Indian Ocean tsunami was that 97 percent of the lives that were saved had been saved before any international aid had even been raised, let alone reached the affected populations.

My experience has reiterated to me that both local structures and provision of relief are fundamentally based on the three forms of social ties. They are not only there before crisis responses, as President Clinton noted with the Indian Ocean tsunami, but also critically shape the effectiveness of those responses. In Japan, for example, the efficiency of waste management after the 2011 Tōhoku earthquake and tsunami has been shown to be directly related to the three types of social ties, bridging, bonding, and linking, and the extent to which they were present between responders, authorities, and communities. In fact, responses to crises are most impactful when the responders have deep and diverse social ties.

Which brings us back full circle to Ros Ereira and Maia Majumder, who we met at the start of this chapter, and their use of social connections to effect change in the face of perhaps the two most complex global crises of recent years.

"A COMPLEX WEAVE"

Remember that Ros had just got a call from the London police force, because the refugee march had topped ten thousand sign-ups. Even in that heightened atmosphere, no one could have anticipated the political outcry that would be sparked by the death of three-year-old Aylan Kurdi, nor the effect this would have on the public support for the march.

It is hard to think of many instances in living memory when a single image changed the way the world perceived an issue. We probably have to go back to the pictures of civil rights demonstrators in the US in the 1960s or anti-imperial India in the 1940s and how particular iconic images shifted public opinion in favor of the campaigners. Or how the 1969 pictures of the Earth from the moon's orbit catalyzed modern environmental movements.

The picture of Aylan, a Syrian Kurdish child whose family had attempted the Mediterranean crossing, his lifeless body washed up and lying facedown on a beach in Turkey, was certainly one of these visceral moments: that single image gave expression to everything that people thought was so wrong about the current situation. The strength of feeling that the picture evoked was remarkable: politicians of all stripes and leanings were now part of a dramatic shift—and in some cases, reversal—in levels of concern and anger about the refugee crisis. The French president called it a reminder of the world's responsibility to refugees. Right-wing media toned down the anti-refugee rhetoric for a whole day and asked what could be done to stop more Aylans losing their lives in this heart-wrenching way.

It was not just the media and the political worlds that had shifted, but also public awareness and conscience. By midday on the Thursday that Aylan's picture went viral, the number of people who had signed up for the march Ros started had leaped to over thirty thousand.

Digital technologies had helped to foster and facilitate a wide-ranging, diverse, and rather unpredictable movement, and did so more quickly than we would ever have been able to do otherwise, and at greater scale. Many other marches started to be launched in London and other major capitals around the world, but Solidarity with Refugees was the biggest, and the flag bearer. It grew and spread like wildfire, coalescing with many other organizations, groups, and networks. Huge international NGOs like Amnesty International and Save the Children threw their weight behind the march, and so too did small local groups.

Farther afield, marches in other capital cities used Solidarity with Refugees as both inspiration and template.

As a result, the march itself became the focus of media interest. Ros wrote a widely shared piece for *The Guardian*. From humble beginnings, the movement she had initiated had captured the public spirit and was seen as representing the shift in the moral landscape. Her click moment had put her at the heart of a dynamic network of turbulent social networks—forming, re-forming, and coalescing, much like the march itself.

It was not just luck and timing that meant that Ros took on the responsibility that she did. She had no money, power, or formal role, but had to ensure complex tasks with multiple responsibilities happened within very tight time frames. She wove together different kinds of networks and relationships to make the whole march happen, even when individuals and whole networks were seemingly in conflict with one another, as is unavoidable in organizing efforts this large. Her honest and authentic leadership and management were essential to the march achieving its goals.

Suddenly it was the day before the march, the weekend before the EU meetings on refugees. We all went to sleep with the same questions: Would people come, or was a social media march little more than an expression of "slacktivism"? Would it be safe and secure? Would we have the impact we wanted? There was nothing to do but go to sleep, but it was far from an easy night. Our worries were unfounded—and how.

The BBC report that following Saturday afternoon is one I have kept. Below an aerial photograph of London are the words "the biggest show of national support for refugees in living memory." The police estimate was that over one hundred thousand people marched, and it was entirely peaceful.

Reflecting on the day, another friend and colleague, Lyndall Stein, a veteran of the anti-apartheid campaign in South Africa, wrote the following astute observation of Ros's Connector qualities:

> When we arrived I noticed . . . an absence of official banners or signage and wondered who had "called" the demo. It was full of families and friendliness and lots of homemade banners. It was also really big: Many thousands marched off to fill Parliament Square for the rally. The mystery deepened until a young woman, Ros Ereira, took to the stage and explained that she had called for the demo on Facebook expecting friends and family to come. When she got 90,000 "likes," she decided to approach the big campaigning groups for help. This is such an interesting change: Instead of Amnesty International or Oxfam inviting an individual to join their demo, Ros was inviting them to join hers. She explained to the rally that she did not belong to any party or organization, but as an independent citizen she had acted—an inspiring demonstration . . . the connections that happen offline are critical to the networks happening online . . . these all create a complex weave, difficult to describe or organize, but a vital part of a powerful trajectory of anger, determination and resistance that needs to remain vigilant, endlessly inventive, responsive and creative.

Despite the obstacles, we did have a fantastic march and great speakers, including the leaders of the Labour, Liberal Democrat, and Green parties, the directors of Amnesty International and Refugee Council, and representatives of refugee groups. Musician Billy Bragg, one of my heroes, came to play as well. We had messages of support from leaders around the world, including Jan Egeland and David Miliband. One of them wrote this to us: "The Solidarity with Refugees march, organized on the internet as a grassroots campaign, is more than just another demonstration. It is a public mobilisation in defence of our shared humanity and our shared values."

Shared values were also very much in evidence in Maia's network. Her COVID-19 hackathon worked. Within a few months, over one hundred researchers were collaborating on twenty-three research projects, looking at everything from predicting the spread of COVID-19 to assessing the quality of media coverage about the pandemic, through to analyzing the collective psychological trauma of its aftermath, with many projects leading to publication in peer-reviewed scientific journals. Maia found that the network was able to bring together people who wouldn't ordinarily have worked together, and that really strong research could result.

Maia highlighted the vital importance of the bridging role of the network in an article for *Wired*: "Without question, the diversity of the network—across disciplines and institutions but demographically, too—has been a tremendous boon to the formulation and investigation of problems that really matter."

This bridging not just of institutions, but also of languages, genders, and religions was especially important for Maia on a personal level and a professional one. The sense of her pride in the network is clear in her voice when she tells me that "two-thirds of the oversight committee were women of color—especially important given that most health practitioners are women, but most health scientists are men." As she put it: "It enabled the kind of networking that is easy to overlook but impossible to overstate."

Just as Ros found, such networks are seldom characterized by complete harmony. Indeed, they work best when there is a degree of constructive friction: a balancing of the forces of competition and cooperation. This was something Maia and her fellow convenors anticipated in the research effort, with two oaths being required of all members.

The primary oath was that each member would commit to keeping the space free of prejudice and open to researchers of all backgrounds.

The second was that topics and issues discussed on the network were to be kept within the network until such a time that instigators of ideas were ready to share them.

Together, these set down the shared values of the network: to be open and participatory, to be multidisciplinary, and to actively support bridges across hierarchies, disciplines, and national boundaries. As Maia told me:

> These oaths enabled really free and open conversation in the network—they really enabled a space that was sufficiently safe . . . where people could trust that not only they will get lots of different perspectives but they will also be able to develop valuable shared assets in that space.

Maia also worked to link network members to established expertise and reputation wherever possible, leveraging her own networks and engaging mentors, supervisors, and former colleagues. For the most part these older, established academics agreed to play a role in the network, partly due to the goodwill Maia herself had generated through her past engagements and collaborations, but also because what the network offered made it easy for them to say yes: a quality paper with a motivated junior researcher and increased credibility because of their involvement.

Perhaps the most touching part of Maia's work is how over time this highly intellectually focused network has led to the emergence of bonding ties: genuine friendships and support in the face of the pandemic. One member noted "a deeper sense of caring emerging from the pandemic's shadow. Conversations about our struggles are now openly shared when, months earlier, they may have been left unsaid."

Over the 2020 Christmas break, when many members could not go home for the holidays, Maia and her fellow oversight members organized a 24–7 Zoom room where network members could join and watch a movie together. As Maia puts it, "These small gestures make a difference . . . folks needed to feel that they were a part of something."

I ask her, does this kind of thing not happen anyway in organizations? Her reply is telling:

> I am immersed in other formal organizational spaces where I am a participant—and I feel that the answer is often no. Institutions always ask for a narrower version of who you are. Ultimately we need to create spaces where people are allowed to feel human . . . and this network is successful because it lets people be their whole selves: they are parents and grandchildren and musicians and artists and chefs and they have the space to discuss that and that is vital.

I got the feeling that while Maia created this network for the benefit of the members, it was also a place where she—and her own remarkable networked diversity—can have a sense of belonging.

As with the refugee campaign networks, Maia's initiative is far from the only example of such networks emerging during COVID-19. As she notes:

> The pandemic is a problem that crosses disciplines. It requires devising plans to reopen the economy while being mindful of public health, or developing strategies to distribute antiviral drugs and vaccines while making sure they're affordable. By forcing public health researchers out from behind the walls of their home institutions and into fully virtual workspaces, the pandemic has in many ways enabled the kind of collaboration that science needs most.

We have seen how the power of Connectors applies to how we harness social ties on a daily basis, in the small-scale daily processes of networking that sustain our lives when we face job or financial or health threats. But as Ros and Maia's efforts show, Connectors are also critical for large-scale and high-profile endeavors that can transform lives and change the world when it is most needed.

Were Paul Revere able to have a cup of tea across time and space with Maia Majumder and Ros Ereira, I suspect they would naturally have a lot to discuss and share. The real trick of Connectors like Revere, Maia, and Ros is ensuring that individuals and teams can connect and bond in dynamic and diverse ways without being overwhelmed by intense information flows and constant changes in social and professional relationships.

That makes their role in forging and facilitating relationships and networks especially important. They have to have high degrees of network intelligence and understanding of when to draw upon different kinds of social ties—bridges, bonds, and links—to effect change.

One of the things they might all marvel at over the course of this time-spanning cuppa is the sheer serendipity that seems to surround their actions. Seemingly momentary psychological processes can, in the minds of Connectors who seem to simply happen to be at the right place at the right time, have remarkable social and political effects.

But the reality is that they weren't just at the right place at the right time.

As we have seen in this chapter, Connectors' good fortune is what happens when preparation meets opportunity.

Effective connection is not merely a leadership capability. In the right settings and contexts, it is the very *essence* of leadership. As Albert-László Barabási puts it, what this means is that we should "replace the corporate ladder with a social bridge."

9

CORROBORATORS

ASSEMBLE THE KNOWLEDGE

SABRINA'S JIGSAW

In the 1987 science fiction horror movie *Predator*, there is a sequence where Arnold Schwarzenegger, playing the lead and last surviving character from his paramilitary rescue team, methodically sets up a series of traps for the eponymous alien. Lost deep in the Central American jungle, Arnold's character, Dutch, fixes taut ropes between creaking tree boughs, sprinkles gunpowder onto large tropical plants, and daubs himself with thick mud to make himself invisible to the alien's infrared vision. Finally, when all the preparations are complete, he lights a torch and lets out a deep, feral howl.

Everything that follows shows how much detail Dutch has put into thinking about the response of the Predator: every action, intention, and sequence of behaviors that he has anticipated and attempted to deal with using the resources of the jungle. As the intricately planned pieces fall into place, we marvel at his foresight while remaining fully gripped by the possibility of success or failure.

About ten years after *Predator* hit the screens, a teenage girl is walking around a derelict building in Newport, Wales, working out the best place to sleep for the night. The setting could hardly be more different: this is a concrete jungle rather than an organic one, but she is using the same precision and forethought as Schwarzenegger's fictional soldier as she starts to manipulate the environment around her chosen spot. And she has the same aim: to stop predators in their tracks.

She stacks up cans of paint next to a doorway so that she can pull them down into the path of any would-be assailant. She piles up newspapers at key points: useful both as obstacles and missiles. Then, when she finally settles down for the night, it is with a heavy wooden plank placed close to her as a weapon.

Sabrina is fifteen. She is homeless. That she is unknowingly emulating the actions of a Hollywood movie character would be of no comfort to her. She was, as she put it years later, "entirely vulnerable. And I knew it. I'd see danger everywhere."

Following the death of her father from brain cancer, Sabrina left home and was living on the streets. She would carry on with school for a further year—with her teachers proving woefully negligent about the state of her life, despite the obvious signals that she was fending for herself with no home to go to at the end of the school day. She carried on trying to learn, carrying her books around with her, and each night, she would have to find somewhere to sleep. After a few unfortunate mishaps, Sabrina started to develop a strategy, based on jigsaws:

> The only way to ensure my safety was to notice everything. All of the time. In my state of hyper-vigilance, I was always looking for another piece of my jigsaw, which I'd test and test and test again. I'd pull it apart and check it once more. It was crucial to do so. My safety depended on it.
>
> I spent a lot of time creating dangerous scenarios in my head and built mental jigsaws to deal with them. Every potential threat

was another piece. I needed to predict horrible situations—and anticipate how things might unfold—so that I could avoid them . . . I learned to second-guess everything and everyone.

But these piecemeal pictures of situations are something we are all accustomed to building. In fact, we do so constantly, to make sense of both the world and our decisions. As Sabrina later put it:

Jigsaws are significant to all aspects of life. Every decision you make and every action you take in your day-to-day existence is based on what you believe you are dealing with. If you start to see things in a particular way, time after time, it becomes your lens. Your way of making sense of the world.

And just like with actual jigsaws, your decision-making brain fills in the gaps. Sometimes it will do this on the basis of evidence, information, and experience. A jigsaw piece you have just carefully put into place has the corner of an ornate building on it, and you might proceed by carefully examining the box, the surrounding pieces, your memory of similar buildings in real life, and finding exactly the next piece that matches to provide the building's capstone.

At other times it will be on the basis of assumptions about what you are seeing. Instead of consulting the evidence available, you are convinced that the building is a church and is topped with a spire, and look for the piece that fits this assumption. Or more often, you will use a combination of your analytical and intuitive skills, meaning that most decision-making jigsaws are a composite, made up of analysis and intuition.

Sabrina's jigsaw, when she was trying to stay safe on the streets, was extreme, and—unlike the fictional Dutch's—all too real. But most of us simply aren't that systematic in times of high pressure. Stress impedes our ability to absorb and make sense of information, and to think and act strategically. Instead, there is a pronounced—even *legitimate*—tendency to act first and think later.

The key to Sabrina's survival on the streets was the realization that her jigsaw "sense lenses" should not be left to chance: "We shouldn't rely on them 100 per cent or leave them unchallenged." While this may be obvious, Sabrina's insight, based on her lived experience on the streets, is that in times of pressure and stress, when we are more likely to just give in to instinct and gut feel, this is *exactly* when we need to be most rigorous.

In following her remarkable journey from the point when she was a homeless teenager, we will learn more about Corroborators, the fifth of our archetypal Upshifters. Their raison d'être is to see beyond the obvious. Whereas most of us tend to act first and think later, corroborators are serious about assessing the viability of ideas, and reaching decisions using logic and critical thinking.

Corroborators are obsessed with understanding the chains of causality that are triggered by actions. The most successful corroborators are people whose capacity for critical thinking does not diminish under stress and pressure, but who find new and creative ways of being rigorous in spite of—or even because of—the pressure they face. To understand more about how Corroborators, we have to return to the neuroscientific foundations of Upshifting that we learned about in part I.

WINDOWS OF TOLERANCE

I was staring at a website on my phone that was offering a world of cuisines and dishes that would arrive at my door in under five clicks and thirty minutes. Embarrassing as it is to recount, I may even have been actually salivating slightly and feeling unreasonably excited at the prospect of paying too much for unhealthy food that would taste half as good as something I could cook myself.

Only a few hours earlier, I had been waxing lyrical about my new healthy-eating regime, and how takeout was a thing of the past: "Why on earth would I want to put that junk into my body?"

Why, indeed. If you're anything like me, this will not be an unfamiliar situation. In the 1980s, in a study that would eventually be cited in the announcement of his 2017 Nobel Prize for Economics, Richard Thaler wrote about a "dual self" in understanding the influence of rationality and emotion on our decisions, and argued that our decisions are influenced by two selves: a *doer* and a *planner*. More than neat metaphors, the two selves have a strong grounding in neuroscience.

"Doer" thinking happens in the evolutionarily older, deeper areas of the brain: the brain stem, cerebellum, and limbic system. These parts of our brains specialize in bottom-up processing—involuntary and unconscious responses to experiences that happen on an emotional and a physiological level. As a result, doer actions are based on impulsive thinking and short-term perspectives, and are driven by desire or the need for instant gratification. Hence, my food delivery antics.

"Planner" thinking happens in the neocortex layers of the brain. More recently evolved in biological terms, this part of our brains is responsible for higher order brain functions—our voluntary and conscious cognitive responses to the world around us. It helps us to pay attention and focus on things; to recall, retain, and revise information; to consciously learn; and to apply information and learning in decisions. It is also needed to control impulsive behaviors and regulate emotions, and is responsible for regulating our response to stress.

Another Nobel laureate, Daniel Kahneman, made the same distinction, saying that doer thinking is "involuntary, extremely robust, and constantly happening outside of conscious awareness" ("think fast"), and planner thinking is "a slower, more effortful decision-making pathway characterized by deliberation and conscious thought" ("think slow").

As my Friday night dinner choices prove, for me, "want" outruns "should" nine times out of ten. This is especially the case if I have had a busy or stressful week, or if I am with friends, or if I have had a drink or two. We all have moments—some of them recurring—when despite our best intentions, our doer brain simply bypasses the planner brain. On the other hand, we may also face situations in which, instead of doing, we overthink. Instead of acting in ways that we are barely conscious of, we are paralyzed into inaction by the sheer weight of our cognitive processes.

The battle between doer and planner underpins much of what is written about and advocated for in enhancing self-control, dealing with compulsive behavior, and overcoming addiction. Self-help books have sold by the millions by offering us all better strategies for dealing with variants of this internal struggle. In reality, as Daniel Kahneman has gone to some lengths to demonstrate, neither the planner nor the doer should have the upper hand, especially when it comes to situations of pressure and stress. Both the planner and the doer are imperfect.

Switching the lens back to Sabrina, in her last two years of school, she was homeless. Being able to go to school was one of the few things that she could rely on—education was the one area of her life where she had some control. This meant carrying around textbooks and school equipment with her wherever she went.

But in one building where she took shelter, a shady character subjected her to a violent, racist attack, and stepped between her and her most prized possessions. He came at her with a broken bottle, and while a few of the other homeless people in the same building, including Sabrina's friend Peter, managed to stall him, her safety was only temporary. Sabrina needed to run, but she couldn't bear to leave her books behind:

I didn't want to run away . . . but I certainly didn't want to be attacked again. I wanted to study, to pass my exams. They were my ticket out of this mess. I needed to get my books. But Dick had a weapon . . .

With hindsight this shouldn't have been a particularly difficult decision. But the weight of the choice was too much. I wasn't just deciding whether to stay or go, but trying to balance all the ramifications, the implications that were inextricably linked.

Fortunately Peter made the decision for me. We left. I'm alive.

As Sabrina found here, sometimes if our planner brain tries to wait for clear and unambiguous information in pressured situations, we can easily miss the opportunity to act. At that point, when she paused, mid-escape, she was contemplating what it would mean to leave her books, and by extension, her education, behind. And so she could escape now, but it could simply be into a bigger trap with even fewer life options.

But at other times our doer brain will fill in a gap with a missing piece based on assumptions or instinct that turns out to be wrong. She could have decided that she was going to fight for her education, gone back, and been fatally wounded.

In an ideal world, we would be able to fill in these gaps in our jigsaws based on direct observation and data. This is the equivalent of poring over the pieces we already have put together, looking at the gaps, and looking at the pieces we still have available. But when large pieces of the puzzle are missing—as they often are in situations of pressure or crises—our brains tend to fill them in. Drawing on assumptions, beliefs, and half-formed thoughts, we use our gut instinct to work out what the missing pieces are.

So what can we do better in such situations? The best answers from the realms of cognitive psychology and neuroscience build on the inverted U of Upshift I shared in the introduction. You will remember that at low levels of stress arousal, we don't experience enough stimulation to be alert and effective. Conversely, at high levels of stress, we cannot focus on the task at hand and become more concerned with the stress itself. In between is the zone of Upshifting, where we have enough stress to keep us on our toes but not so much that it becomes overwhelming.

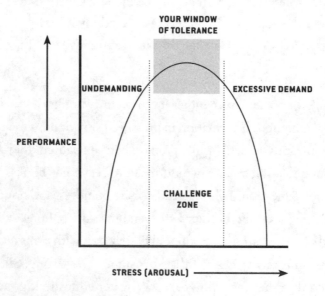

Stress and trauma specialist Elizabeth A. Stanley has described this zone—the Upshifting zone—as our neurobiological window of tolerance. This is the space within which we are capable of regulating our stress levels upward or downward to remain within optimal performance under pressure.

The reason for this is simple. It is in this Upshifting zone that we are most likely to effectively integrate the doer and the planner brain processes. Inside the window the two parts of our brains "work together as allies." And the people who do this best under pressure are Corroborators.

Evidence has shown that the doer and planner parts of our brains react differently across the Upshift curve. The doer process works at all levels of stress, but actually increases as stress does, as our survival instincts kick in more strongly. This is in part, incidentally, why our memories of stressful situations can be so patchy. The planner does best in the middle zone but falls away at high stress levels. Interestingly, a well-known child therapist has used Keith Jarrett's Cologne concert and the broken piano as an analogy for the window of tolerance:

> The fact that Jarrett had to keep within the mid-range of the keyboard, the playable range, while avoiding the unplayable lower and upper registers of the piano brings to mind the notion of the window of tolerance and the importance of enabling children to stay within their own emotionally tolerable mid-range. The upper register of Jarrett's piano was harsh and tinny, the lower range deep, rumbling and inaudible; an aural evocation of . . . the child who is either too chaotic or frozen to playfully engage, or the dissociative child who has learned to survive through disconnection.

Outside the window, by contrast, the planner and doer parts of our brains are more adversarial. This manifests in a number of ways that I'm sure for most of us are all too recognizable.

Degraded thinking brain leads to poor situational awareness, anxiety-based planning, defensive reasoning, distraction, and memory problems.

Doer hijacking is when our emotions and stress bias perceptions absorb attention and drive decisions and behavior. In such settings, our planner might even become subordinate to our doer, and work to justify the actions ("I deserve a takeout, I've worked really hard this week").

And finally, *thinking override* is when we retreat into our heads and get disconnected from gut instinct, emotions, and physical signals. This can lead to suppression, compartmentalization, ignoring the reality of the situation, and just pushing on regardless.

But within their window of tolerance, Corroborators are more likely to do the following:

OBSERVE, GATHER DATA, AND POSE QUESTIONS	• Perceive internal and external cues about events and situations • Obtain and absorb adequate and appropriate information • Objectively assess and integrate that information • Accurately assess a specific phenomenon as an opportunity or threat
PROBE, EXPLORE, AND TEST APPROACHES	• Search for all possible options in terms of actions • Evaluate each action in terms of costs and benefits • Compare options systematically through experiments • Plan and consider their likely future effects
MAKE CHOICES, LEARN FROM THEM, AND MAKE CHANGES	• Choose the strategically optimal action • Align actions with values and goals • Consciously assess the ongoing consequences of chosen actions • Adapt and learn for the future • Advocate for necessary changes

How we operate within and outside our windows of tolerance can have a profound effect not just on specific situations but on our entire lives. Indeed, some psychologists have suggested that our entire personalities, behaviors, career paths, and even choice of partners and friends are determined by the conflict and the collaboration between the planner and the doer systems in our heads.

As Stanley puts it:

> The wider an individual's window, the more likely that they can keep online their capacity for . . . effective integration of the [planner] brain and [doer] brain . . . even during levels of high stress and intensity.

This was clearly true of the young Sabrina. Aged sixteen, in the face of insurmountable challenges and shocking indifference from teachers, and having sat her exams, she left school. Her hoped-for route out of homelessness had changed, and things became desperate:

> When you live that life, you feel invisible. You feel like a ghost in society. If someone in the street falls over, people rush over to help, but there you are on the street corner with no food in your belly, nowhere to live, no clean clothes and people walk past you like you are not there.

Sabrina had to change tack—and the answer came not from her planner brain or her doer brain but a fusion of the two. Her mentality shifted: her jigsaws became less about how to survive, and more about how to live. She observed, probed, asked questions, and made changes. Her new path out of her desperate situation was not school, but from enterprise.

> I thought, "This can't be me. I can't stay in this life." I started selling the Big Issue [a UK street newspaper set up to give homeless people income opportunities] but competition in Newport was fierce—I'd make about £15 a day at best . . . so I'd get a bus to Monmouth every morning and work from 7 a.m.–7 p.m. until I'd sold every copy. It took a few attempts to get off the streets and into secure accommodation . . . Eventually, I scraped together enough money to put down a deposit on a tiny rented

flat in a deprived area of South Wales called Risca—a town far enough away that I wouldn't be recognised. No one would look at me with pity; it was a fresh start.

And the implications of this fresh start were very clear. She describes the moment that her window of tolerance widened: "This was the first place where I could close the door and feel safe. For the first time I wasn't just using tactics to survive, I was beginning to form a strategy to live." As she says, "That's when I started to think about what else I could do with my life."

But before we find out what her answers were, I want to turn to another remarkable young woman who was asking very similar questions about her life's purpose, albeit a century and a half earlier.

"VERY LITTLE CAN BE DONE UNDER THE SPIRIT OF FEAR"

When Florence Nightingale was a child, she loved collecting, organizing, and analyzing things. Shells, plants, butterflies: they would all be gathered and meticulously documented on reams of paper. By age six, in 1826, she was keeping a record of her prayers, to compare in systematic form what had been granted and what had not, and thereby calculate the efficacy of prayer.

Contrary to expectations placed on girls at that time, she showed a keen desire and interest in subjects such as mathematics and physics, and went out of her way to get tuition in them, despite her mother's stern disapproval. Unusually named after the splendid Italian medieval city in which she was born, and which would become a popular girls' name thanks largely to the fame she would gain, parental efforts to steer her toward the more refined interests of a young lady's education via a European tour backfired somewhat.

Each city they visited led not to heightened cultural and artistic appreciation but to long and detailed diary entries documenting population statistics of each capital, and the size and location of hospitals and charitable organizations providing health and social services to the populations. But more than this, she began to note the effect of politics on the everyday lives of ordinary human beings. Whether prisoners, soldiers, poor women, beggars, or chambermaids, she talked to everyone and documented what they told her. Suffice it to say, she found politics deeply wanting as a way of organizing human lives, writing in her journal: "What a system to exist in 1832!" While her sister filled her notebook with paintings and drawings of famous monuments and replicas of artworks by the Old Masters, she remained steadfastly and firmly uncultured, much to her mother's dismay, deigning only to draw a single small urn in the margins of her notebook.

When she eventually decided upon a profession, this choice, too, went against her parents' wishes: nursing was considered lowly and disreputable for a woman of her standing. But she was nothing if not determined, repeatedly saying, "God has called me to serve." She noted the exact date that her calling took place: February 7, 1837. It would take a few more years for her to work out exactly how she would serve.

There are two paintings left to us of her most famous and influential professional experiences in the 1850s, when still in her thirties. The first shows her standing in the middle of a large busy room, holding a large sheet of paper, while next to her is a uniformed man gesticulating in a manner that seems to indicate either annoyance or distress at the information contained on the paper. The second, more famous picture shows her surrounded by bandaged men in beds, holding a lit lamp. Although she became known far and wide as the Lady with the Lamp, thanks to this second picture, Nightingale could equally well have been known as the Lady with the Lists and Tables. It would have been much more accurate a quality she would certainly have appreciated.

When Nightingale arrived at the British military hospital in Turkey in 1856 along with forty female colleagues, what she encountered was a disaster within a crisis. The Crimean War had a religious pretext—about the rights of two different Christian minorities to access religious sites in the Holy Land—but at its heart was an old-fashioned land and power grab.

The two relevant churches—Russian Orthodox and Roman Catholic—had in fact worked out an agreement, but regardless, the leaders in France and Russia were determined to go to war anyway. The conflict that ensued, which would pull in many countries across Europe, would become infamous for its "notoriously incompetent international butchery," not least because it was one of the first that used not just modern military technology, such as explosive shells, but also one that was extensively documented and reported thanks to the rise of photography and telegraphy.

Of particular note for the British was the disparity between the British and French levels of care, as lamented by William Russell, lead writer for *The Times* of London:

> Are there no devoted women among us, able and willing to go forth to minister to the sick and suffering soldiers of the East in the hospitals of Scutari? Are none of the daughters of England, at this extreme hour of need, ready for such a work of mercy? Must we fall so far below the French in self-sacrifice and devotedness?

When Nightingale arrived at the Scutari hospital with her team of nurses, it was thanks to a direct commission from Sidney Herbert, the secretary of war. Soldiers and doctors alike were suffering from collective trauma and it showed most clearly in the hospital, where she found filth, carnage, and chaos. In January and February 1854, three thousand soldiers died of dysentery, frostbite, and gangrene, in four miles of beds arranged eighteen inches apart.

As superintendent in charge of women nurses, Nightingale set about establishing much-needed order and method in the hospital. Initially her efforts were limited to traditionally female household roles: food and clothing. But in fact, this enabled her to get at two issues she saw as fundamental: nutrition and hygiene.

She soon became known as the "lady-in-chief," and a big part of the reason for this was that she had clear arguments, backed by data and evidence. She started out by doing what came naturally to her: asking people what they thought. Having got the soldiers' perspectives, she moved on to establishing detailed lists and records on food and hygiene. This met with a great deal of resistance. For example, on dietary records, she struggled to change the doctors' custom of filling out diet sheets for the following day with what the patients *should* have been given to eat—despite advocating for the importance of accurately recording what they were actually given. Indeed, many doctors claimed such a change was a physical impossibility—because the diet sheets for the current day had already been filled out, and the only sheet they had was a blank one for the next day.

Nightingale's thirst for data, analysis, and mathematical results bemused the doctors, who saw statistics as "useful for hard-pressed officials who wanted to prove their points against assailants, but useless for practical men who merely wanted to cure the sick." Those who took notes did so primarily in order to conciliate Nightingale herself. Moreover, "during the time of pressure all such notetaking was perforce abandoned." The lack of British statistics were her ongoing despair, as she wrote to the London War Office authorities: "The Medical Statistics . . . are in a state of great confusion, so that it is hardly possible to obtain correct results."

Her frustration led her to engage with other doctors and hospitals, where the French facilities in particular filled her with "admiration, envy and amazement." One French army doctor led her from the tents of the sick to his own tent to show her his cahiers, "complete records of every case, devoting to each individual patient not merely a statement of diet but also daily notes of medical observations." The unhappy contrast with the British doctors did not stop there, as this description of the French medical attitude conveys: "The [British] doctors, like the politicians and the clergy, failed to see that right action can only be based on correct thinking, and that only on free inquiry and careful attention to results."

This was as self-evident to her as it was that inquiry and thought were useless unless they led to action. It was the combination of the cahiers with an intense energy and humane attitude of mind that she admired in the French. This spirit of fusing rationality with emotion was one that she aspired to herself, as she put it: "I think one's feelings waste themselves in words; they ought all to be distilled into actions which bring results."

Now a world-famous—if reluctant—celebrity, feted by royalty and politicians, as well as an adoring public, Nightingale did not become any more of a socialite than she was before Crimea. Instead, she set about using statistics to advocate for a wholly new approach to military medicine. "She wanted data for each regiment, monthly, on mean strength with the mean number of sick from fever, consumption and other causes, then aggregated to the whole army." In her view this would allow the secretary of state for the War Department to "see the movements of the health of the army as clearly as the movements of time on the face of a clock."

Only with this complete system of statistics would "we be able to administer the laws of health with that certainty with which we know they are capable of being administered." But data alone was not enough. Tables and diagrams litter the pages of her notes. Nightingale was an innovator not just in medicine and in statistics, but also in how information about both could best be conveyed. And here she showed a deft understanding of appealing to both rationality and intuition.

Whether her childhood love of charts helped here is uncertain, but she developed a range of different types of graphs, akin to what we today would understand as pie charts, with which to bring dense statistical findings to visual life. The most famous of these was the coxcomb.

Unlike a pie chart, in which the areas of the slices of the pie represent the proportions of the data, the length of the slice—measured from the center—also conveys data.

The coxcomb below shows how all the causes of mortality were greatly reduced during her second year in Crimea (the left-hand chart) compared with the first (the right-hand chart).

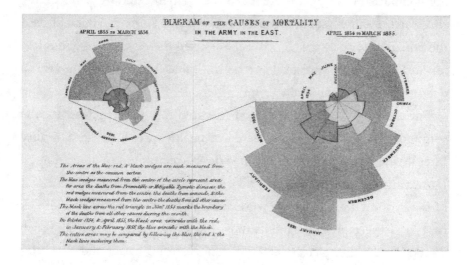

In her coxcomb for illustrating causes of mortality in Scutari during the first two years of her mission, the chart was divided evenly into twelve slices representing months of the year, with the shaded area of each month's slice proportional to the death rate that month. Her color-coded shading indicated the cause of death in each area of the diagram. By breaking down the data month by month she was able to show instantly what would not have been read in tables or paragraphs. Her ambition, as she wrote with a mixture of exasperation and hope, was to "affect thro' the Eyes what we may fail to convey to the brains of the public through their word-proof ears."

It worked. In the 1850s, the British had one of the worst records in military medicine in terms of excess unwarranted deaths. By the late

1860s, it was widely acknowledged to be the best in the world. And a major reason was the Corroborator zeal that Nightingale brought to her postwar work.

Part of the reason she was such an advocate for statistics was that she was deeply aware of their limitations and spent years studying how to navigate them. In particular she was aware of the challenges of drawing comparisons across different types of patients and of using simple singular measures (such as mortality) without sufficient context, and the potential for data manipulation. She used these skills to good effect not just to advance statistical corroboration as a discipline, but also to debunk weak or poor studies.

For example, one study seemed to show that nursing was not the solution that Nightingale advocated in health care, because mortality rates in one hospital had not dropped despite employing trained nurses. However, Nightingale argued and demonstrated that this was a result of faulty thinking: the nurses were deployed to help the most complex cases and therefore the patients most likely to die. A better approach would have been not to selectively report, to do better and more detailed comparisons of patients who received care and those who did not, and not to focus on singular measures (in this case, mortality).

She would go on to advocate for the wholesale transformation of all medical practices with statistics, arguing that "improved statistics would tell us more of the relative value of particular operations and modes of treatment than we have any means of ascertaining at present . . . and the truth thus ascertained would enable us to save life and suffering, and to improve the treatment and management of the sick."

In this, she would prove to be a prophet, although it would take almost another century and another brutal conflict for her ideas to be realized. In 1941, a young Scottish doctor named Archie Cochrane would apply a statistical approach used predominantly among agricultural researchers to work out what treatments would be most effective in treating prisoners of war in Greece. The randomized control trials that he would advocate, building on the principles that Nightingale set out, have become the mainstay of evidence-based medicine.

People like Nightingale, and like the French doctors she so admired, are able to maintain and even widen their windows of tolerance during crises, pressure, and stress. But such things come at a personal cost. Of all the Upshifting practices, this would seem to be the one with the harshest payback. It takes a special type of energy to maintain our logical and intuitive minds in the Upshifting zone for long, as we will see next.

Nightingale understood, like all Upshifters, that "very little can be done under the spirit of fear." But defying that spirit was something that she paid for with a lifetime of debilitating illness: fatigue, stress, and anxiety. As she would later put it, "There is no part of my life, upon which I can look back without pain."

Nightingale would remain a devoutly religious person her whole life. She became more religious as she got older, keeping her Bible and prayer book always close at hand, and praying daily. Later in life, she would reflect on mathematical analysis as being akin to a religious exercise, aimed at the discovery of codified laws in the social sphere. She wrote, "To understand God's thoughts we must study statistics, for these are the measure of His purpose." Her six-year-old self, carefully charting the efficacy of prayers, would no doubt have agreed.

MIND THE GAP

In recent years, the highly demanding and stressful nature of our lives has led to many of us experiencing slow-burning mental health challenges. If that weren't enough, the rise of COVID-19 in 2020 was enough to put the entire world into a state of perpetual heightened anxiety. Mindfulness and meditation have gained a lot of credence as ways of dealing with mental ailments without resorting to, or as a complement to, medical treatment.

This is best demonstrated with the growth of the market for mindfulness training courses and apps dedicated to supporting our well-being: in 2018, over half of employers in the US offered some form of mindfulness training to workers. In January 2021, mindfulness apps were estimated to be a multibillion-dollar global industry and set to quadruple in the next few years. Of course, this has come with a degree of backlash and cynicism amount those who see the movement as "hippie-dippie nonsense gone global."

In fact, much of this wealth of training and learning materials has been informed by the work of a handful of leading academic scholars, chief among them University of Massachusetts professor Jon Kabat-Zinn. Working initially with chronic pain sufferers in the 1970s, Kabat-Zinn developed an eight-week course to teach patients the basics of mindfulness meditation, which he defines as "the awareness that arises from paying attention, on purpose, in the present moment and non-judgmentally." For those who aren't familiar, mindfulness focuses attention on breathing and allows meditators to become more aware of their bodies and minds as they change moment to moment.

The course proved popular and was soon being offered to people after industrial accidents, cancer patients, paraplegics, and those suffering from depression. What was especially striking was the evidence base: study after study showed the benefits of mindfulness extend far beyond what many "hippie-dippie" accusers would be willing to accept. It enables chronic pain sufferers to better deal with their symptoms, including lessening their subjective experience of pain. You are up to a third less likely to relapse into depression if mindfulness is part of your treatment. You are also more likely to be able to maintain your focus and attention on complex tasks if you regularly practice mindfulness meditation.

Kabat-Zinn's original course has been applied in adapted form to many different professions. Most interesting for us is how valuable it has proved for people working in extreme and pressured situations.

Elizabeth Stanley, whose work on widening the window of tolerance we learned about earlier in this chapter, has been one of the main exponents of this. A former soldier herself, she has used Kabat-Zinn's work as the basis of a program she calls "Mindfulness-based Mind Fitness Training." The US Department of Defense has made this a core part of pre-deployment training for soldiers, and has also funded rigorous neuroscientific and psychological studies into its impact on combat troops.

One of the common findings of pre-deployment research is that standard training does not in fact help soldiers: it leads to higher levels of anxiety and stress and declines in cognitive performance. Stanley's mindfulness training has been shown to have a marked benefit. Participating solders demonstrated greater cognitive capacity—and in particular were more able to maintain the planner-doer balance under stress. They reported lower stress levels before deployment, and, even more remarkable, their stress arousal was much more efficient before and after combat drills: they would be more likely to reach peak performance during drills, and they would also recover quicker to a calm state afterward. All of this has been measured using bioharnesses that the soldiers wore before during and after drills as well as by analyzing blood samples. At a psychological level, the self-reports of the soldiers who had undergone the training showed that they were much more able to see the combat situations as challenges as opposed to threats. Mindfulness, far from being a hippie pursuit, can in fact help soldiers operate in the Upshifting zone.

In a similar study of firefighters who were provided both relaxation and mindfulness training, Stanley's long-time collaborator Amisha Jha explored how exactly mindfulness helps how people experience, deal with, and navigate stress. She found that firefighters use a combination of planning and doing: at times, slower, effortful, and deliberately controlled; and at other times, fast, automatic, and emotionally charged.

Thanks to mindfulness training, the firefighters became less reactive and less prone to impulsive actions. At its best, mindfulness helped the firefighters to make their jigsaws more explicit and therefore improve them.

Rather than simply strengthening their inner planner at the expense of their doer, the mindful firefighters became better self-observers *during* incident responses. At a neurological level, they were able to press pause on those automatic doer pathways in their brain that were created by prior experience and triggered by stress. This meant that they could observe the actual moment-to-moment inputs of incidents as they happened, interrogate them, and integrate them into their decisions. As a result they were more likely to come up with adaptive responses to stress and pressure, and to think through the implications of possible responses through thought experiments and scenarios.

Being better at interrogating their actions and feelings using data and evidence yielded benefits in their personal lives too, with many reporting higher levels of well-being and improved personal relationships. They were also more likely to recover faster after stressful situations—not just in specific situations, but also when suffering conditions of prolonged stress such as PTSD.

Elizabeth Stanley argues that mindfulness training is one of the best ways of getting what she calls the thinking brain and the survival brain to "work as allies." But she also suggests that the mindfulness industry, with its focus on rapid results, is mis-selling both the practice and its benefits. In a curious reversal of expectations, she has advocated an almost Zen-like approch to mindfulness: don't aim for immediate wins, expect to get worse before you get better, practice consistently, and have faith that you are cultivating valuable capabilities. Less self-help, her ideas are redolent of how super-elite athletes we saw in part I achieve peak performance.

This, incidentally, gives a fascinating new interpretation of Florence Nightingale's use of daily prayer: through meditative contemplation— mindfulness, in modern parlance—she was using the best way she knew to recover from the intensity of her experiences during the Crimean War.

At its most effective, mindfulness broadens the range of situations in which we are able to integrate our doer and planner brains, and widens our window of tolerance. And it can do this even in the most extreme of situations.

Which brings us back to Sabrina.

SABRINA'S CHOICES

When we last saw her, Sabrina had recently moved in to her own flat, and was reflecting on what she should be doing with her life.

> I knew what it felt like to live nearly every day like it's the worst day of your life so I wanted to do something to help people in a similar situation. I think, in a funny kind of way, I wanted to rescue other people in a way that no one had been able to rescue me.

The solution was to turn to a profession where she would be judged not on the basis of qualifications but "on the strength of who they believed I could be." Risca, the small Welsh town where she settled, had a part-time fire station. She applied to join the fire service, passed the interview, and became the station's first female firefighter. Promotions and postings around the UK followed.

And then a dramatic incident changed her career trajectory. Sabrina's crew was called to a scene where she knew her fiancé's team was responding and a firefighter was reported to have been badly burned. Desperately hoping it wasn't her fiancé, Sabrina simultaneously felt guilty and relieved when it turned out to be a different crew member.

Wracked by guilt, Sabrina spent hours trying to understand why such accidents happen. She drew many of the same conclusions that Captain Sully had found in his own work on safety and accidents: that most accidents were down to human error.

This led her to launch a new research group at Cardiff University, focusing on incident command decision-making. For her the fundamental issue was "how we operate when we make decisions that ultimately affect whether people live or die."

It turned out that the fire service was far behind other high-risk sectors such as the military and aviation in terms of even considering this issue. Sabrina, inspired in part by her father's slow and debilitating illness and death from brain cancer, was especially keen to contribute to neuroscience. She was fascinated by what was going on in her brain, and the brains of her firefighting colleagues, at a neural level when making life-or-death decisions. Before her own work, there had been very little real-time study on these decisions. Much of what was understood came from interviews and analyses undertaken *after* incidents, which were subject to distortions and biases.

What Sabrina did, along with one of her colleagues, was to strap video cameras to incident commanders' helmets so entire responses could be recorded, and to use this footage as the basis of interviews and research. This proved useful not just as a means of uncovering what had actually happened during incidents, but also provided a level of self-awareness to incident commanders, who were able to adjust their behavior as a result of seeing the real-time footage.

What this research revealed was that despite the assumption that commanders used highly analytical and rational processes (in planner mode) the reality was that this applied only around 20 percent of the time during a firefight. The other 80 percent of the time, the decisions commanders made were based on gut feelings (in doer mode). Sabrina became especially interested in how firefighters could better balance what she saw as the planner and doer when responding to an emergency:

> I didn't simply want to describe how decisions were being made and to find the potential traps. I wanted to help commanders make better decisions . . . I wanted commanders to prompt their un-

conscious consciously and question how their actions link to their goals. I wanted them to open up their jigsaws.

This led her to develop the Decision Control Process—a framework that acknowledged how the brain actually worked and could help people optimize their decisions. While mindfulness might get to such a place indirectly, the Decision Control Process could provide a step-by-step methodology that could be used in training and assessments. In essence, whenever a commander has to make a decision—whether analytical, intuitive, or a combination—they are prompted to ask themselves three questions:

1. Goal: What do I want this decision to achieve? This links the decision to overall aims of a response.
2. Expectations: What do I expect to happen as a result? This increases situational awareness, and doing this as a group develops the shared jigsaw across the response team.
3. Risks versus benefits: How do the benefits outweigh the risks? This ensures better, more effective and robust decisions.

After numerous trials, Sabrina and her colleagues found that the approach worked: commanders linked their actions to their plans, and situational awareness increased significantly. Most importantly, the technique did not slow down decision-making. It is now being used to inform all UK emergency decision-making principles and codes, and has been taken up around the world including in the US, Europe, Australia, and Hong Kong.

Thanks to this simple and powerful approach Sabrina developed for evaluating options and making decisions in fast-paced and unpredictable situations, she has become an acknowledged expert in some of the most intensely pressured situations human beings face. Along the way she also gained a PhD in behavioral neuroscience from Cardiff University. Her ongoing research examines the high-pressure situations firefighters regularly face to improve life-and-death decision-making on the front line.

But Sabrina acknowledges that she is only ever as good as her capacity to corroborate. As she puts it:

> The techniques we have developed though our research have contributed to making firefighting, and keeping firefighters, safe . . . [but] they only work if we practice them all of the time . . . All because we were determined to change and believed in constant and active critique . . . to challenge the status quo of what you think you know, peel back the layers of assumptions, and find new perspectives, new angles and new ideas.

As we have seen at numerous points in this chapter, corroborating exerts a considerable mental toll. Sabrina's honesty on this is illuminating:

> While I'm not as intense these days, some of those old habits are ingrained in me. I'm never satisfied with the picture I have. There are times when this can be a powerful asset . . . however there are other times when I have to make a conscious effort to stop myself before it becomes all-consuming. I still have a tendency to overthink things and assume the worst. It makes me unnecessarily anxious.

Today, Dr. Sabrina Cohen-Hatton has become the youngest chief fire officer in the UK, with a PhD in the neuroscience of decision-making. Her ability to make rigorous decisions under pressure transformed not just her own life but also how emergency services respond to crises in the UK and around the world. She is an Upshifter par excellence.

Beat that, Arnie.

10

CONDUCTORS

ORCHESTRATE CHANGE

RANDOLPH

I want to introduce you to someone who, perhaps more than any one person, was the catalyst for this book and the ideas within it.

Randolph Kent has had an illustrious career in the United Nations, operating on the front lines in some of the very worst crises in the world: Ethiopia, Sudan, Rwanda, Kosovo, Somalia. I started working with him in London in 2004 after a chance encounter in an airport—thanks to the bridging networks of a mutual colleague—after which he recruited me to be his deputy on an exciting new R&D program on humanitarian work. The project was an attempt to reform aid responses to global crises—to try to make them fit for the twenty-first century. There was something about it that sparked my imagination—and still does.

Perhaps it was because his mission emerged from abject failure. Randolph had led the United Nations humanitarian effort in Rwanda in the aftermath of the genocide in which eight hundred thousand people were massacred between April and June 1994. A lot of aid was given—from the cynical perspective, this was the governments of the world giving "too little, too late" having not done more to stop the genocide when it was actually happening.

What Randolph realized in leading this effort was that so much of that aid was to meet the material needs of the people affected and displaced by the horror. Almost nothing was being done to deal with the psychosocial needs of the population. As he put it in a recent interview:

> When a nation has witnessed the death of over 800,000 of its citizens in three months, there in retrospect was a clear need to provide psychological and social-psychological assistance . . . I should have been more sensitive to the need for new types of requirements, I eventually began to recognize the fact that we had to think differently, in a more anticipatory and adaptive way.

What emerged from this Upshift was a dedicated effort to bring about wholesale changes in the way the sector operated: how humanitarian responders plan, deliver, and even think. For anyone in the know working in the sector today, Randolph's fingerprints are to be seen everywhere: in new initiatives, new policies, even speeches by senior leaders.

Back in 2004, when I first met him, I was curious to learn more about how he had led the humanitarian work in some of the worst crises in the world. His reputation and networks were astonishing—but I wanted to get to the nitty-gritty of what he actually did.

But from the outset of our professional relationship and friendship, he was always incredibly reticent to talk about his own experiences. He somehow, through charm and disarm, always managed to turn every conversation around so that he did more asking and listening and probing of me than talking about himself. Although I found this both amusing and slightly frustrating, I would come to realize that this was in fact the heart of the way he worked—not just in meeting rooms and conferences, but on the front lines of crises. You will have some sense of the measure of the man that I learned much more about his considerable achievements by observing and talking to others. In fact—not for lack of trying—it took almost twenty years for him to open up to me about his experiences.

When other colleagues talked about him, the most common description was along the lines of "Randolph is the best master of ceremonies the humanitarian sector has ever had." But we all saw something more than this too—he was not just a skilled emcee.

We learned in chapter 3 about how William Rutherford and Kate O'Hanlon worked to bring many different professions together—from hospital porters to nurses to brain surgeons—to transform how the Royal Victoria operated during the Troubles. Their very presence made a difference to those around them. And this was exactly what Randolph brought to the humanitarian sector. People like this play one of the most vital parts in the orchestra of Upshifters: they are the Conductors of the whole show.

What set Randolph, the Royal Victoria leaders, and others we will learn about in this chapter apart is not how they Upshift themselves, but that their primary way of doing so is to create the conditions and enabling environment for others to Upshift. I would get a clearer view of Randolph's Conductor role in the aftermath of the Rwandan genocide from an unexpected source.

The international investigation into the response to the horrors of Rwanda was searing in its criticism of the foreign policy, diplomacy, peacekeeping, and aid efforts. The catastrophic political and military response was compounded by the failings of the humanitarian agencies, which the report found to be chaotic, uncoordinated, and underprepared. It found that the aid system had a "hollow core" where there should have been strong leadership. And yet, one of the chairs of the investigation told me many years later, there were some bright lights—and Randolph was one of them. In fact, his successes only highlighted the profound failure of the rest of the aid system.

And it was in these successes that I found the embryonic ideas that would inform the whole of this book, including the last of the Upshifter archetypes I want to share.

As well as being a wonderful mentor and colleague, Randolph, more than anything, helped me believe in myself. Over the years, I met so many people in the sector who felt the same way about him—including many in very prominent positions. They helped me realize one of the truest qualities of any leader is not how they shore up power, but how they share it. Randolph was the kind of leader who created leaders.

And he did this even in the dire conditions of post-genocide Rwanda. One of his junior colleagues had continued to specialize in conflict and peace-building work after the genocide, working across Africa and Asia. I met him fifteen years after he had worked with Randolph for just a handful of months, and we found ourselves talking about our mutual mentor and connection. He told me, in a way that I found both moving and illuminating: "Even today, I ask myself at crunch points or crises, what would Randolph do?" I think this is an extraordinary articulation of how a leader can help to instill an enduring challenge mentality in those around them.

In the time I have known him, he has been a champion for creative ideas in the sector, providing space, support, and airtime for many approaches that have gone on to transform the way aid work is done. In fact, there are a whole host of creative solutions that Randolph encouraged and supported during his time in Rwanda that have left their mark on the sector to this day.

He was the first person to have the role of United Nations humanitarian coordinator. There was no written job description, and while he was formally the coordinator, the reality was that the rapidly assembled aid agencies did what they did without regard to others working around them—and certainly not to a UN operative with a jumped-up job title.

An immediate problem arose with the information he had available. There was no coherent data on what was being done by whom. When he asked the humanitarian responders to pass information through him to their HQs, the typical response was a polite but firm no.

After a meeting with NGOs on information sharing, Randolph had to travel to Nairobi, Kenya, in the back of a cargo plane, leaning against sacks of grain, thinking about how to crack the information problem. He came up with the idea of a common tool that all the NGOs would use to provide information on the "who, what, where" of the crisis response. The key here was making it a shared regional network. While this will not seem so groundbreaking today, bear in mind this was in the days when the internet was on the horizon as "the next big thing."

Once it was set up, the information started to flow in from a few sources. Here was Randolph's insight: that he could use the inherent competitiveness of the aid responders to covertly get them to collaborate (in his words, "gently encourage through subtle stimulus"). When the idea of the network was shared more widely, those not contributing to the platform were all too aware of their absence. So too was everyone else. The initial blockers realized that they had to be on the platform or risk being invisible. And soon enough, the platform became a thriving exchange for information about the unfolding crisis. Today that regional information network has expanded and evolved into the leading humanitarian media platform in the world.

But the most remarkable thing that he achieved in that terrible situation was to get all of the decentralized and competing parts of the aid system to work together with a sense of shared purpose. After a major violent incident at one of the biggest camps for displaced people in the country, in which many were killed and injured, Randolph and his team were in charge of the aftermath, to get the camp back in order.

Using the style that I and his other colleagues would come to instantly recognize—asking, listening, probing, orchestrating—he was able to get all of the elements of the international system working together: logistics, identification of injured and survivors, and delivery of much-needed medical aid to where it was most needed. As he described it to me, he was aware that he "was at the core, bringing all of this together."

In fact, the chair of the post-genocide investigation attested to exactly this in no uncertain terms: "Before Randolph, the system was hollow. There was no core. While he was there, it worked, somehow, miraculously. He brought things and people together. And after he left, it was hollow again."

The measure of his success was made clear to Randolph when the head of the Rwandan branch of the International Committee of the Red Cross—the closest we have in the humanitarian sector to US Navy SEALs special ops: enigmatic, highly insular, and extremely effective—came to him in the midst of the chaos and drama, and asked, "How can we help?"

As he tells me this story, it is clear that these four words remain to this day one of the greatest accomplishments of his professional life. And to me, it speaks volumes about Randolph's role as a Conductor of collective purpose.

It may be unsurprising to learn that Randolph vigorously dismisses that there was anything special about what he did: no grand plan, no vision, just a way of approaching, listening to, and engaging with people.

So what was it that underpinned his performance as a Conductor in these profoundly challenging circumstances? He thinks and then says simply: "Perhaps it was the intense pressure."

COVERT LEADERSHIP

In the 1990s, management scholar Henry Mintzberg undertook what might have seemed like a curious experiment. Long aware of the metaphor of the orchestra conductor to describe the act of management and leadership, he decided he would actually spend some time observing and watching how conductors worked and what they did, both during and in between performances.

Before he had the idea of stepping into the rehearsal room, he looked at the work of twenty-nine nonmusical leaders, in very different settings. He spent a day with the head of the NHS in the UK. He observed the work of a Red Cross manager overseeing refugee camps in Tanzania. After accumulating a wealth of data and information about leadership, in his words, "I found the idea of spending a day with an orchestra conductor irresistible." He was so struck by what he found that it became the mainstay of a widely cited *Harvard Business Review* article on how to manage and lead professionals.

Mintzberg studied Bramwell Tovey, conductor of the Winnipeg Symphony Orchestra, over a number of days. What he found was that the orchestra was "structured around the work of highly trained individuals who know what they have to do and just do it." They were, in fact, like many organizations of professionals—he made the comparison with a consulting firm or a hospital.

Mintzberg concluded, using a phrase coined by Bramwell Tovey, that "in such environments, *covert leadership may matter more than overt leadership*" [emphasis added].

Covert leadership is worlds apart from the image of control that many people still associate with leadership. But it certainly echoes what we have already seen with how Randolph orchestrated crisis responders together. In the broader context of Upshift, Conductors are people who lead without seeming to, neither authoritarian nor completely powerless.

Mintzberg concluded that there was indeed much for leaders to learn from conductors: it was not about control, showmanship, and ego—but rather "act[ing] quietly and unobtrusively in order to exact not obedience but inspired performance."

In 2015, Danish composer and conductor Ture Larsen set out to test this. Drawing on his experience, he found that there were three things that conductors did for their orchestras that echoed strongly what Mintzberg saw in the work of Bramwell Tovey.

1. They provide courage to deal with stress and anxiety (mentality).

First and foremost, conductors support the players and instill *courage in the face of stress, pressure, and crisis*. One famous opera conductor, Mark Wigglesworth, puts it as follows: "[The conductor] needs to send a message of complete confidence to the musicians, some of whom might quite understandably be feeling nerves of their own. The goal is to be able to [operate] with both conviction and trust. Adrenaline and oxytocin in perfect equilibrium." Conductors enable and coach challenge mentalities.

2. They create space and encouragement to improvise in the face of challenges (originality).

Conductors build a *shared sense of creativity and originality* by listening and observing, finding pockets of people, and bringing them together in a shared interpretation of the music, and by also giving them the space to learn and adapt. The best conductors strike "a balance between giving guidelines and allowing musicians the freedom to play and express themselves."

3. From their unique vantage point they provide a big-picture overview of the challenge and enable shared solutions (purpose).

Last, and by no means least, conductors create a sense of *shared purpose* and help to integrate many different goals within this, enabling coherent and original performances based on a keen awareness of the technical and human capabilities they bring together.

Explaining this in Upshift terms, every single player in an orchestra is a Corroborator, analyzing visual and auditory signals and cues to make sense of what is happening and play the shared score. The best soloists will be the Crafters, working tirelessly to perfect their practice, finding new ways to interpret and perform. Connectors might be the principal brass player who knows and understands the roles of the string and woodwind sections, and therefore are the "bridges" in the ebb and flow of symphonic movements. And the Conductor is focused on the whole of the orchestra, sets the pace and rhythm, and makes the experience a singular, unforgettable one for the audience.

All of this is at odds with the popular image of the conductor on the podium, in complete control of the situation. Instead of bending the orchestra to their will, the best conductors orchestrate the diversity of skills and capacities to create a single shared movement. Larsen's experience as an orchestra conductor was the inspiration. Unbeknownst to him, he had distilled into his course what we can now recognize as the key ingredients of Upshifting. But the most surprising thing was what he chose to do with this knowledge: get groups of crisis responders together and teaching them to conduct "Frère Jacques."

MUSICAL EMERGENCIES—OR HOW CONDUCTORS REFRAME STRESS AND PRESSURE

One of the hardest things to teach in the field of emergency management is how to lead in a crisis. No matter how many simulations and drills new staff participate in, there is little that genuinely prepares them for their first experience of responsibility in the chaos and fog of crisis. Creative approaches to training abound, many of them making use of games, simulations, and, increasingly, virtual technology. But it is hard to refute that crises are themselves the best classrooms for inexperienced emergency responders.

However, there is also a growing sense among experienced responders that there are analogous situations that could benefit early learners, where the same skills could be developed in a psychologically and physically safe space—one without trauma or blood.

When he identified the contributions of conductors to orchestras, Ture Larsen was convinced that music could be this safe space for crisis responders. In collaboration with intensive care surgeons at Copenhagen University Hospital, and building on the latest research on orchestral conducting and leadership, including Mintzberg's work, Larsen worked with clinical faculty and staff to develop a crisis leadership skills course for medical students, nurses, and other health practitioners. What he found in doing so clearly surprised him.

He looked back over thirty years of research into leadership in emergency medical situations, and found that a lack of leadership fundamentally shaped outcomes for people in critical conditions. There was also a range of studies that indicated the leadership qualities needed: calmness under pressure, the ability to "radiate" a can-do mentality, and being goal oriented.

But when he dug further, there was almost nothing that had been done to develop these skills and qualities for those on the front lines of medical emergencies. Instead, virtually all of the work up until that point had been on *measuring* leadership skills: on proving them rather than improving them.

Larsen saw potential for trainees to develop their skills through a novel approach: essentially a musical version of the elite athletes' and freedivers' stress innoculation that we saw back in chapter 1.

All of his training courses were captured on video and made available online—and they make for intriguing viewing. One in particular stands out because of how powerfully it shows a particular participant's mentality toward stress and pressure being reframed.

What follows is the conversation, which I have transcribed and edited from the subtitled exchange between Larsen and a third-year medical student named Jeanett. My own observations are interspersed.

LARSEN: Is it so frightening?

JEANETT: YES!

LARSEN: That is not what you show at all, not at all. Hard to believe.

JEANETT: I think it's horrible!

LARSEN: Shaking?

JEANETT: Yes.

At this point Jeanett bends forward from the waist, so her head drops below the lectern—it is literally like she is being doubled over with stress.

LARSEN: Step aside and have a little time out. When you're okay, come back and do it again. We're starting to get under your skin. We won't get any deeper than this. It's like we dig deep into your soul.

JEANETT: Yes, a lot! Because it's a crazy challenge!

RANDI: Why do you think it is such a great challenge?

JEANETT: I am really, REALLY shy, and so I hate to stand up in front of people. It's so uncomfortable. *[almost in tears]*

Here you can literally see the reframing happen—first at Ture's suggestion.

LARSEN: You know what? I'm shy too. Really! Definitely the guy in the back of the school row. No need to shine up there.

Kind of funny I became a conductor, right? But what I do is learned . . . If you can just focus on putting the first sound of "Frère Jacques." Right there! Then forget yourself.

And this is Jeanett's click moment.

JEANETT: It's very . . . another go!

LARSEN: What I'm doing is learned, but my trick, it's the music. I forget about how I look when I meet the orchestra. I'm focused, right there. Then forget yourself.

JEANETT *[BACK AT THE LECTERN, UNDER HER BREATH]*: Fake it till you make it.

Jeanett takes another turn of conducting, with a visibly different stance and confidence, and finishes to enthusiastic applause.

JEANETT: Right, I wish everyone could try this.

LARSEN: Want to have another go, now you know how much fun it is?

JEANETT *[PAUSES AS SHE IS ABOUT TO SIT DOWN]*: Yes!

She turns and runs back to the podium to the sound of her fellow medics' applause.

It's worth reflecting on what the above scene tells us. Jeanett, as a medical student, was clearly expected to have a certain level of personal and leadership qualities. By putting her in a situation where there was no expectation for her to excel, she was able to actually talk about how standing in front of people and trying to conduct them felt ("It's horrible!"). But by listening to Larsen's guidance on how he dealt with such situations, and reflecting on her own experiences, she was able to reframe the threat as a challenge, and take on the leadership role. Importantly, the fact that the task of conducting was physical rather than verbal meant that the change was not just something she talked about: You could literally see it in her body language and her stance when she had the click moment. She became an embodiment of the eustress phenomena: in the zone. Jeanett was not the only participant to experience such a transformation. In fact, Larsen observed similar changes among most of the participants.

Although this was a simple simulation involving only a piano and fellow medics singing "Frère Jacques," the results spoke for themselves. Many of the trainees reported gaining new insights into how to manage emergency situations. And in the evaluation of the course, the researchers found that "conductors' skills improved and profoundly changed the participating students', nurses' and residents' behaviour." This was thanks to the orchestral methods used to overcome their own anxiety and demonstrate calmness and authority to support those they were leading to do the same. They were better able to change others' mentalities by conducting their way through pressure and crisis.

PITCH PERFECT—OR HOW CONDUCTORS FOSTER ORIGINALITY

On the surface, professional soccer teams and classical orchestras have very little in common. The chanting and cheering of crowds in stadiums around the world could hardly be more different from the rarefied grandeur of concert halls and their audiences. Orchestra players are all on the same team—at least in ideal circumstances—and their efforts seldom if ever make front-page news. Footballers play together but also compete with other teams in every theater of action, and their efforts are among the most assiduously followed of any group in the public eye. Musicianship continues ripening into old age, whereas soccer players' careers are all too brief, lasting around eighteen years at most.

Bramwell Tovey, the conductor we met earlier who coined the phrase "covert leadership," also memorably described himself as akin to a soccer coach "who also plays." He is far from the only person working in one of the two fields to have drawn such parallels. At a talk on leadership at Harvard Business School, legendary Manchester United coach Sir Alex Ferguson said that he felt a profound kinship while watching an opera: "I had never been to a classical concert in my life. But I am watching this and thinking about the co-ordination and the teamwork—one starts and one stops, just fantastic. I spoke to my players about the orchestra—how [they] are a perfect team."

And here is Jürgen Klopp, the coach of Liverpool FC: "A football team is like an orchestra, you have different people for different instruments, and some of them are louder than others, but they are all important for the rhythm."

This sense of similarity goes both ways. Sir Simon Rattle, conductor of the London Symphony Orchestra (LSO), has said on more than one occasion that he is akin to a soccer manager. In fact, he once described himself as Jürgen Klopp, "just with hand gestures. If he could do what he did with that team, but on the spur of the moment . . . just with hand gestures, you would have an idea of what our job [as a conductor] is. It is very clear what an extraordinary difference he made to that group of people."

And Rattle would know all about making a difference to a team. When he took over at the LSO, it was in the spirit of a returning prodigal son, having spent sixteen years at the Berlin Philharmonic. The LSO was in the doldrums at the time—so used to the absence of the previous conductor that players learned to conduct just to be on the safe side—and often sounding "coarse and unpleasant." In contrast, during Rattle's tenure they are widely acknowledged to have a "new spring in their fingers."

Because of my interest in Conductors, I was curious to explore this connection further. Are there deeper similarities between the two professions?

One obvious thing was the popular perception of coaches and conductors themselves. While clearly crucial for success of the teams they lead, these two jobs are also universally misunderstood and underappreciated. Rattle was recently the subject of a documentary where he was repeatedly asked, "But what do conductors *actually* do?" And while all soccer fans bemoan players' miskicks and mistakes, they reserve their true opprobrium for managerial decisions that run counter to all apparent logic and sense, and whose mistakes they would clearly have been able to correct were they to have had the top job themselves.

Beyond this, scientific research has revealed some more substantive similarities and common ground between these two very different kinds of leaders and their players.

The first thing that struck me is how similarly the two groups are described by cognitive scientists. Take these accounts of collaborative playing in two studies: Which do you think describes orchestras, and which soccer players?

> An individual's own actions in performance may be formed as a result of adapting and synchronizing to any number of their colleagues, forming a rapid chain of action and reaction within a complex web of allegiances and hierarchies.

> Performance is characterized by rapid, complex, and coordinated task behavior, and where [players] dynamically adapt proactively and reactively to the environments within which they operate. Coordination between team members becomes therefore critical, considering how the team as an entity dynamically solve[s] tasks.

The first is, in fact, a description of orchestra musicians and the second of soccer players, but I think they are so similar in what they convey about the two groups that they are interchangeable.

The next similarity relates to the experience of flow. The work of Mihaly Csikszentmihalyi introduced this idea to popular audiences, and it has been popularly known as "in the zone." This is how he described it: "Being completely involved in an activity for its own sake. The ego falls away. Time flies. Every action, movement, and thought follows inevitably from the previous one, like playing jazz. Your whole being is involved, and you're using your skills to the utmost." Ture Larsen, in delivering his course for emergency medics, found that almost all of the participants reached individual and collective flow states during the training.

Studies of top-flight soccer teams in the Netherlands, Germany, and Slovakia point to the importance of "group flow" in understanding performance. For many soccer players, this collective state emerges, Upshift style, from the pressure of match situations. In fact, the researchers describe flow among soccer players as the shared experience of eustress, and linked it to collectively moving to the top of the inverted U of the Yerkes-Dodson curve that we learned about in chapter 1. Cross-country evidence suggests that reaching and maintaining shared flow states in match settings strongly correlates with the likelihood of wins or draws for the team in question.

Research on orchestral players has also identified collective flow, defined as a collective "peak experience, a group performing at its top level of ability." This was seen as a "strikingly useful framework" for understanding the processes involved in expert orchestral performance. Indeed, the first scholar to find group-based versions of flow did so thanks to observations of the performances of jazz musicians.

As with soccer players, orchestral flow is also linked to pressure and stress. Elite musicians have described performance as a collective response to pressure, of reaching new heights thanks to the stressful performance situations they had to work with as a matter of routine. As one musician put it: "Everyone is taking a risk because music is all the better for these extremes." This is very reminiscent of Keith Jarrett taking the risk to do the unexpected with a broken piano. But instead of one man jamming by himself, this is an entire orchestra collectively riding the waves.

But it is not just the collective flow state that unites soccer and orchestra players. The things that get them to this desire state are also identical: how the members of the team or orchestra relate to one another, and how they relate to their coach or conductor.

For soccer players, relating to one another is described as "shared situational awareness"—the common interpretation of cues or the overlap of each member's individual level of situational awareness, which allows for action that is both accurate and expected by teammates. This isn't merely cognitive but is physical too: technology has allowed scientists to analyze how individual players move their heads during gameplay and has shown that higher frequencies of head movements—linked to exploring the environment around them—are positively related to the individual player's game performance. For the soccer players studied, this shared awareness was one of the most important determinants of whether teams would achieve a flow state in gameplay.

This exact same phenomenon was also critical for musicians in achieving peak performances—although they described it differently. Elite symphony players often describe the most important set of skills in achieving excellence as being akin to "radar": "Listening to, communicating with, and adapting to other members of the ensemble at all times during rehearsal and performance." At a certain level of achievement, technical playing skills are the basic entry requirement. Radar is the quality that makes the difference for highest-level orchestral work.

> Radar [is] a collective, conscious process that arguably develops in subtlety as the familiarity between [players] increases over time . . . [This is] an internal process which is imperceptible to those on the "outside," involving collective action in order to

align with prominent parts within the ensemble for the greater good of the performance.

As with soccer players' shared situational awareness, radar was seen by players as a "vital component for achieving satisfying and high-quality orchestral section playing, acting as a mechanism for facilitating spontaneity of musical expression at a collective level"—in other words, it was the route to collective originality.

But linking to other players was not enough. The coaches and conductors were also integral to achieving collective flow. Shared situational awareness and radar are seen most often and are most effective when accompanied by collective guidance. In soccer, this is also known as the coach's "master plan." Indeed, research suggest that this relationship with the coach is the *primary means* by which a soccer squad grows into being a team rather than a group of individuals. When coaches are unable to guide and structure team experiences to facilitate shared situational awareness and coordinative and adaptive action, the result is ineffective team performance. This form of leadership promotes flow and enables creative team working. One study even showed that it can can also help ward off serious injuries.

In orchestras, too, conductors' communication styles, personalities, and approaches fundamentally shape how musicians interact with one another, establish shared radar, and experience creative flow. At an operational level, this boils down to listening, attentiveness, and awareness. At a more strategic level, there is the conductor's equivalent of the coach's game plan, which is developed during rehearsal (the equivalent of training for soccer players) and is manifested in live performances. The best conductors understand and work with the music and harness their musicians' creativity in such a way that the orchestra performs in a unified and original way for the audience.

Let's step back for a moment and think about Conductors more generally. The latest evidence on leadership in times of pressure and stress—much of which has focused on the response to COVID-19— argues that the fundamental requirement is for leaders to establish a collective "window of tolerance." They do this by themselves operating within the Upshift zone at the top of the inverted U. And they also enable those around then to do so, through a process described as "positive emotional contagion." The best Conductors—Rattle, Klopp, Randolph, Sister O'Hanlon—*positively* infect those around them, listening to them, hearing their perspectives and views, and making them comfortable "exploring, learning, innovating, making mistakes and growing." Because this is a fundamentally human capability, any one of the other Upshift archetypes can, with time and practice, learn to be a Conductor, to be in the center and balance the diversity around them.

Conductors aren't only to be found in the realms of specific professional disciplines or pursuits. Some of the most influential have used positive social contagion to make profound changes to the way our society as a whole works and how it values people.

DO EVERYTHING—HOW CONDUCTORS CATALYZE SHARED PURPOSE

The year 1816 has gone down in European history as the "year without a summer," thanks to the cataclysmic explosion of the Tambora volcano, on the island of Sumbawa in modern-day Indonesia.

The aftereffects of Tambora were far greater than the temporary grounding of air travel the world saw in 2010 after the eruption of the Icelandic volcano Eyjafjallajökull. Imagine if the clouds had lasted for months—in a persistent fog that would not be dispersed by rain or wind. Imagine if global temperatures fell as a result of the lack of sunlight. The eruption of 1816 was not a disruptive event. It was devastation.

To this day, the 1810s remain the coldest decade on Earth since records began. There was snow in New York in June. Rivers and lakes and even the sea in some places froze over. Harvests failed everywhere. People went hungry and died in the tens of thousands. And the lack of oats meant hundreds of thousands of horses died.

In an improbable consequence of the eruption, the crisis also triggered one of the most important social changes humanity has ever seen. And the unlikely catalyst was the bicycle.

The widespread loss of horses led a German civil servant to experiment with alternative means of personalized transportation. The year 1817 saw the development of the very first velocipede—the earliest known ancestor of the bicycle, sans pedals. Think of a large-scale version of the balance bikes that small kids use today when learning to ride. Karl Drais was a brilliant Crafter—and prolific too. He also came up with the first-ever typewriter with a keyboard, the first-ever meat grinder, and a foot-powered railway cart still used today.

The velocipede became popular, as did its variations. But there was a problem. Drais's Crafter mentality only took his invention so far. On his first ride, he traveled seven kilometers on what was widely known as the best road in his locale. But most roads were pitted and pocked, and the swathe of velocipede riders who followed in his wake in the years after had to swerve and sway dramatically to avoid being upended. So they shifted onto pavements, posing a considerable danger to pedestrians—and as a result velocipedes were banned across Europe.

Decades later, Drais's invention came to new life and popularity, with the introduction of pedals and safety measures, including brakes. By the 1860s, they were still seen as something of a risky contraption. It was with the development of the safety bicycle in the 1880s that the true transformative effects of the bicycle would be seen. And this was not just because of what it could do, but also for whom.

Previously only seen as companion riders on two-seater machines,

Victorian era women took to the safety bicycle in their masses, to the consternation of many men at the time.

This happened thanks to changes at a remarkable number of levels, all happening at the same time. It may be hard to imagine today, but we would probably have to leap forward to the 1960s and the introduction of the pill to see an equivalent transformation in gender norms thanks to a single technological advance.

Many kinds of Upshifters—beyond the designers of the bicycle itself—got on board. Women's clothing had to be adapted by an army of ready Crafters to allow for easy and safe riding, leading to growing popularity of shorter skirts, divided dresses, the notorious bloomers, and even trousers.

The blurring of the Victorian symbols of masculinity and femininity did not go down well. Doctors started to suggest that bicycles led to women looking flushed and unwell rather than pale and ladylike. Many argued about the sexual connotations of the bicycle: the idea that women might be able to feel sexual pleasure while riding was shocking to the mores of the day.

At the same time, defiant groups of women made the Challenger counterargument about the positive effects—on women's strength, fitness, and mental ability. The great American women's rights campaigner Susan Anthony wrote in 1886: "Let me tell you what I think of bicycling. I think it has done more to emancipate women than anything else in the world. I stand and rejoice every time I see a woman ride by on a wheel."

Corroborators around the world started to write about the benefits of cycling—in women's magazines, church publications, even in temperance societies.

In a remarkably short time the bicycle was changing Western society's ideas of acceptable femininity. This movement was heavily influenced by a number of Conductors who saw an opportunity in the humble bicycle to enable significant liberation for women. They orchestrated the efforts of enthusiastic riders, clothing and fashion designers, women's magazine publishers, employees' unions, legislators, and even paving companies to achieve empowerment. The goal was simple: accessible, cheap freedom of movement.

These Conductors oiled the wheels of interaction and collaboration. They brought together people with different ideas, opinions, and life experiences, spurring innovation—what economists sometimes call "social agglomeration."

While there are many fantastic examples in this mass-distributed movement, one person in particular played a vital Conductor role. She did so by adapting to the physical innovation of the bicycle, showing how it could be embraced without fear and used to change every facet of her society.

Her name was Frances Willard. Before taking up the bicycle she was *already* considered one of the most famous women in the United States, thanks to her work on reforming many aspects of the United States constitution. She led the Women's Christian Temperance Union, which would become the largest women's membership organization in the world under her presidency. And not only did Willard start riding a bicycle at the relatively late age of fifty-three, but she also wrote a book about it.

As one scholar has put it, Willard was all too aware that "the bicycle craze—with manufacturers, commentators, designers, and riders among its relevant actors—changed the conditions within which women were living their lives that could not be accomplished through argument alone."

Willard was not merely seeing the bicycle as a reflection of social change, but saw as cycling women as the active creators and shapers of the new arguments that would enable social change to happen. She was well aware of her own covert role as a Conductor within this broader process of change: "A reform advances most rapidly by indirection . . . an ounce of practice is worth a ton of theory."

Willard was sensitive to the changes that were happening around her, and positioned herself carefully at the interface between technologies, social norms and attitudes, and commercial practices. She was also all too aware of the fragility of these changes, and the importance of courage among women to be able to capitalize on them. As another scholar puts it:

> Simply in learning to ride the bicycle herself, and then describing her experience in a memoir, Willard was demonstrating her awareness of the vulnerability . . . publishing her actions so that other women might, collectively, be inspired to enact the change with their bodies, and "re-signify the social order" by riding the bicycle.

I have an old and worn copy of Willard's book, which sets out in wonderful detail each stage of her learning, from early practice runs during which two young men had to assist her to later stages when she would race her friends around public squares. Illustrated with photographs throughout, the most powerful part for me is this passage:

> Learn on a low machine, but "fly high" when once you have mastered it as you have much more power over the wheels and can get up better speed with a less expenditure of force when you are above the instrument than when you are at the back of it. *And remember, this is as true of the world as of the wheel* [emphasis added].

The phrase "the New Woman" was coined in 1894 by novelist Sarah Grand as a way of describing the new approach and mentality of the women of the decade, as distinctive from their less-assertive foremothers. Not only did Willard encourage women to take up bicycles, but she was also clearly positioning them as an instrument of social change.

Here she is, describing how she ran a previous campaign for social change as a diverse orchestra:

> Twenty years will have elapsed since the call of battle sounded its bugle note . . . This was the one wailing note . . . playing on one string. It caught the universal ear and set the key of that mighty orchestra, organised with so much toil and hardship, in which the tender and exalted strain of the Crusade violin still soars aloft, but upborne now by the clanging cornets of science, the deep trombones of legislation, and the thunderous drums of politics and parties.

And here she is again, advocating the "Do Everything" policy, which is a wonderful articulation of Conductors as "all-round advocates":

> A one-sided movement makes one-sided advocates. Virtues, like hounds, hunt in packs . . . An all-round movement can only be carried forward by all-round advocates; a scientific age requires the study of every subject in its correlations. It was once supposed that light, heat, and electricity were wholly separate entities; it is now believed and practically proved that they are but different modes of motion. There is no better motto for the "Do Everything Policy," than this which we are saying by our deeds.

Many young women responded to her call, with many self-identifying themselves with the label of "Revolting Daughters." Think of a young woman in her twenties in the 1890s: able to get a loan for the first time, paying for a bicycle without any requirement for parental authority, adjusting her skirts, and then getting on her bicycle and riding in whichever direction she wanted, without a chaperone.

Beatrice Grimshaw was one of these young women, and these are her words:

I am a Victorian. I was born in the Seventies, in a big, lonely country house five miles—a whole hour's journey—from Belfast. I was governessed and schooled and colleged. I was taught to ride and play games.

I was taught to behave. To write notes for Mamma. To do the flowers. To be polite but not too polite, to Young Gentlemen. To accept flowers, sweets and books from them, but no more. To rise swiftly with the rest of the six daughters and sons when Papa came into the breakfast-room, to kiss him ceremoniously, and rush to wait upon him. He liked it, and we liked it. I went to dances, and waltzed to "The Blue Danube," "Sweethearts," and "Estudiantina." I went to afternoon parties. I was chaperoned. My three sisters were good girls, and content.

But I was the Revolting Daughter—as they called them then.

I bought a bicycle, with difficulty. I rode it unchaperoned, miles and miles beyond the limits possible to the soberly trotting horses.

The world opened before me.

And as soon as my twenty-first birthday dawned, I went away from home, to see what the world might give to daughters who revolted.

What it gave me first was the offer of a journalistic post. There were maps of far-away places, maps with tantalizing blanks in them; maps of the huge Pacific, colored an entrancing blue. I swore that I would go there. I made a London newspaper commission me; I went.

History doesn't divulge, sadly, whether Beatrice had read any of Willard's writings to inspire her to "do everything." But it is certainly easy to imagine the Conductor's words echoing in her ears as she went into the entrancing blue with revolting purpose: *"And remember, this is as true of the world as of the wheel."*

RANDOLPH'S CODA

Think about the accounts of Conductors that I've described throughout this chapter and this book.

Time and again the same message comes up: it is their ability to bring different kinds of people together—and do so in a way that manages their fear and their risk, harnesses their creativity and originality, and defines and delivers shared purpose—that makes all the difference.

After I presented my ideas and insights to him, Randolph concurred on this general argument but demurred on his specific talent as a Conductor, protesting:

> I wish it was a strategy—but really it was about fortuitous social combinations, taking advantage of something that happened, looking for an alternative way to do things with the people I could work with . . . I wasn't [Napoleon] Bonaparte . . . there was nothing grand about it . . . I was looking for the opportunities for people to collaborate and deliver and just kept bumping into them . . . I recognized when I was bumping into something— and then I used it to good advantage . . . I was fairly innovative, sure, but I have to impress upon you how serendipitous so much of this was.

Randolph's humility is touching, but it also feels a little like deflection. There is a pattern to his work. After Rwanda, the same things were said about him in Kosovo, where he was in charge of bringing the system together after the NATO "humanitarian war." And also in Somalia, where he ran the entire UN operation in the most war-torn country in the world at that time.

If you can make conducting a habit in some of the toughest places in the world, and in ways that few others are able to replicate, surely this speaks to something more than just luck? I think so, and I think it is something that the best orchestra leaders and soccer coaches in the world would understand very well.

That I owe him a huge debt should be clear. He has been a mentor and cheerleader throughout my own career in the humanitarian sector. But he also very much was the original inspiration for this book. Randolph is the first person I ever recognized as an Upshifter. Without his questions and encouragement, his listening and discussion, I would never have started to look at the phenomenon of his own leadership and how he brought radically different types of people together to enact change.

In shining a small light on just a select few of his achievements, I hope to have told a credible story that he would recognize and endorse. But I also think I have learned more about why he is so reticent to do so himself. In his humility, Randolph has always been reluctant to see or describe himself as the story.

I believe that Randolph's Upshift story is that he harnessed, strengthened, and wove together everyone else's stories to harmonize and magnify their collective effect.

And there is perhaps no better description of a Conductor than this.

EPILOGUE

THE MASTER SWITCH

My explorations have shown me that Upshifting is a phenomenon that we can actively learn to switch into, to facilitate new ways of thinking and behaving, and produce results we could otherwise never have imagined. In the opening chapter, I described my own initial experiences of learning to apply the phenomenon of the "master switch" that freedivers exploit to consciously harness their bodies and minds to survive enormous depths on a single breath.

Each of the Upshifters we have encountered in this book metaphorically entered new waters with each challenge they faced, and made conscious choices to fight the stress response and swim, rather than sink.

No matter who we are, we each have the Upshift master switch within us, and we can make more use of it every single day. We can all benefit from reframing the problems we face through an Upshifting lens, and by making more use of the practices and habits that will help us flip the switch.

Ultimately, *Upshift* is an affirmation of our innate abilities and power to enact change, even in the complex and uncertain world around us.

But by changing the way we look at stress, uncertainty can become the very catalyst to create the world we want to live in. As Nobel laureate Ilya Prigogine eloquently put it, our world may be uncertain, but that uncertainty is at the very heart of human creativity.

This applies as much to the small *c* creativity by which we can find a better way to get to work on days when transportation is shut down or harness our networks to deal with job insecurity as it does to the big *C* creativity of ingenious approaches to malnutrition, disease, and even human survival itself. Historian Arnold Toynbee was surely talking about this when he argued that all our great advances represent a creative response to a problem: having read this book, we might call this the "challenge and Upshift" view of the sweep of human history and progress. We stand on the shoulders of Upshifters.

But if I have learned anything over the course of the discovering and documenting of *Upshift*, it is that uncertainty, pressure, and stress are not enough. Necessity is not alone the mother of invention.

We can change how we manage ourselves to open more doors to more possibilities.

We can reboot our ways of thinking to help us walk through those doors.

But having the *mentality* of performance under pressure alone doesn't lead to Upshifting.

We also have harnessing our *originality* to transform our performance. This is about shifting from doing things right to doing the right things. This demands an openness to creative and innovative practices, and a credible vision of how to move the needle from how the world *is* to how it *could be*.

In *Upshift*, I hope to have also shown that we can all contribute to moving this needle, whether as Challengers, Crafters, Combiners, Connectors, Corroborators, or Conductors. We can play to our strengths and learn to overcome our weaknesses, either by ourselves or even better in tandem with others.

But *originality* is also not quite enough to Upshift.

I finished this book in the middle of the third UK lockdown following the 2020 pandemic. Unlike the first, when I was working around the clock on the global humanitarian response to COVID-19, my son Koby is home-schooling. As I write these words, he is working on some art next to me: carefully drawing abstract patterns onto the leaves of an alien plant that resulted from his afternoon's imaginings.

It's a strange feeling, watching him: somewhere between calm and concern, holding the space between anxious and devil-may-care. For many of us, the pandemic has meant doing this day after day after day, holding ourselves in a permanently alert position instead of relaxing or moving. It is like holding the "master switch" moment of an Upshift in perpetuity. It is agonizing at times, but there is no choice. I have to do it, for him and for me.

And that is perhaps the single most important thing I want to share as I bring this book to a close—the most important thing uniting every single Upshift I have had the good fortune, preparation, and timing to learn about and experience.

You have to have a *purpose*: a reason to start Upshifting, and a reason to carry on doing it.

Sometimes our sense of purpose centers on things that are surprising and unexpected. Sometimes it is self-evident. I think of my mother that day on the barge, as I chatted away to the armed soldiers. I look across at Koby, his tousled dark chocolate hair falling over his face, as he carefully inscribes a pattern on the protruding leaf of his plant, tumbling headfirst into his lush imagination.

Whatever your purpose might be, it centers on the things that make the ingredients of Upshift come together for you.

These are the things that make you realize you have a master switch in the first place.

The things that make you reach out for it.

You feel its contours, and ever so gently, and then with growing confidence, push. And as you do, you feel yourself shift into a new and different state of being and possibility.

One that says that even in the face of pressure and stress, there is a way out.

And it is up.

Click.

ACKNOWLEDGMENTS

Upshift has benefited greatly from the support of an amazing team working across the United States and the United Kingdom: Bryn Clark, Ruben Reyes, Meghan Houser, Bob Miller, and all at Flatiron; Shoaib Rokadiya and all at William Collins. James Middleton and Martin Toseland provided fantastic editorial guidance and support at the early stages and to get over the finish line respectively. Special thanks are due to Jenny Heller, my agent, without whose passion, wisdom, guidance, patience—and forgiveness—this book would quite simply not exist. Never underestimate the things that can happen while waiting for Sainsbury's to open on a cold Tuesday morning.

NOTES

All notes are linked to the first few words of the relevant sentences and paragraphs. The body of the note sets out the sources, and, where relevant, how they have been used. This system has been chosen to avoid overloading the main text with superscript reference numbers.

FRONT MATTER GENERAL

ix *For Upshift definitions:* These definitions synthesize Dictionary.com definitions of *downshift* and *upshift* together with material synthesized from readings on stress, pressure, and performance from a psychological and evolutionary neuroscience perspective, including: Fabienne Aust et al., "The Relationship Between Flow Experience and Burnout Symptoms: A Systematic Review," *International Journal of Environmental Research and Public Health* 19, no. 7 (2022): 3865; and K. G. Bailey, G. Cory, and R. Gardner, "Upshifting and Downshifting the Triune Brain: Roles in Individual and Social Pathology," *The Evolutionary Neuroethology of Paul MacLean: Convergences and Frontiers*, 2002, 318–43.

INTRODUCTION

7 *a wealth of research:* Francesco Montani et al., "Examining the Inverted U-Shaped Relationship Between Workload and Innovative Work Behavior: The Role of Work Engagement and Mindfulness," *Human Relations* 73, no. 1 (2020): 59–93; Carmen Sandi, "Stress and Cognition," *Wiley Interdisciplinary Reviews: Cognitive Science* 4, no. 3 (2013): 245–61; Adrian Hase et al., "The Relationship Between Challenge and Threat States and Performance: A Systematic Review,"

Sport, Exercise, and Performance Psychology 8, no. 2 (2019): 123; and Jenny J. W. Liu et al., "The Efficacy of Stress Reappraisal Interventions on Stress Responsivity: A Meta-Analysis and Systematic Review of Existing Evidence," *PLoS One* 14, no. 2 (2019): e0212854.

8 In the UK alone: Health and Safety Executive, *Health and Safety at Work: Summary Statistics for Great Britain 2018*, October 2018, https://www.hse.gov.uk/statistics/overall/hssh1718.pdf.

8 The Midlife in the United States study: Joseph G. Grzywacz, Dikla Segel-Karpas, and Margie E. Lachman, "Workplace Exposures and Cognitive Function During Adulthood: Evidence from National Survey of Midlife Development and the O*NET," *Journal of Occupational and Environmental Medicine/American College of Occupational and Environmental Medicine* 58, no. 6 (2016): 535.

8 In 1908: Robert Mearns Yerkes and John D. Dodson, "The Relation of Strength of Stimulus to Rapidity of Habit-Formation," *Journal of Comparative Neurology* 18 (1908): 459–82.

10 what psychologists have called eustress: Hans Selye, "Stress Without Distress," in *Psychopathology of Human Adaptation*, ed. George Serben (Boston: Springer, 1976), 137–46; and G. Brulé and R. Morgan, "Working with Stress: Can We Turn Distress into Eustress," *Journal of Neuropsychology & Stress Management* 3, no. 4 (2018): 1–3.

10 Mentality, Originality, Purpose: Each of these areas is discussed in detail in chapters 1, 2, and 3, where sources will be cited.

14 At 1524 Eastern Time: There are many articles, books, and papers on the Miracle on the Hudson—the most rigorous source material is the report produced by the National Transportation Safety Board: D. A. Hersman, C. A. Hart, and R. L. Sumwalt, *Loss of Thrust in Both Engines After Encountering a Flock of Birds and Subsequent Ditching on the Hudson River: Accident Report NTSB/AAR-10/03* (Washington, DC: National Transportation Safety Board, 2010), https://www.ntsb.gov/investigations/accidentreports/reports/aar1003.pdf.

14 What follows is a simplified extract: This is extracted from the Federal Aviation Administration transcript of the interactions between the New York terminal traffic controllers and US1549 pilots. None of the words have been edited, but some of the peripheral communications have been removed to focus on the exchanges between Sullenberger and the controller. The transcripts and the audio recordings can be accessed on the FAA website: https://www.faa.gov/data_research/accident_incident/2009-01-15/.

16 In one of the countless interviews: Unless otherwise stated, all of the quotes that follow are from Captain Sullenberger's 2012 interview with the trade newspaper *Industrial Safety and Hygiene News* at the October 2012 National Safety Council Congress and Expo: "ISHN Exclusive Interview at NSC with Captain Chesley 'Sully' Sullenberger III (Part 2)," *Industrial Safety and Hygiene News*, October 31, 2012, https://www.ishn.com/articles/94407-ishn-exclusive-interview-at-nsc-with-captain-chesley-sully-sullenberger-iii-part-2.

17 In another interview, he spoke: "'Sully' Sullenberger Remembers the Miracle on the Hudson," *Newsweek Special Edition: Amazing Miracles*, September 11, 2015, accessed July 4, 2022, https://www.newsweek.com/miracle-hudson-343489.

17 As he expanded in another interview: MSNBC.com. "Transcript: Captain 'Sully' Sullenberger: My Aircraft," MSNBC.com, August 5, 2020, accessed July 4, 2022, https://www.msnbc.com/podcast/captain-sully-sullenberger-my-aircraft-n1235862.

19 As Abraham Lincoln famously put it: Abraham Lincoln, "Annual Message to Congress—Concluding Remarks: Washington, D.C., December 1, 1862," Abraham Lincoln Online, updated 2020, https://www.abrahamlincolnonline.org/lincoln/speeches/congress.htm.

19 In his classic book: Mihaly Csikszentmihalyi, *Flow: The Psychology of Happiness* (London: Random House, 2013), 200.

1. THE UPSHIFTING MENTALITY

21 The Upshifting Mentality: Two BBC programs provide invaluable source material—the first a documentary in the *For One Night Only* series by Paul Gambaccini: Paul Gambaccini, "Keith Jarrett: The Cologne Concert," December 29, 2011, in *For One Night Only*, series 6, radio program, BBC Radio 4, https://www.bbc.co.uk/programmes/b0103z8j. The second is an account that draws on the experiences of Vera Brandes: Vera Brandes, "Keith Jarrett in Cologne," November 5, 2011, in *Witness History*, radio program, BBC News World Service, https://www.bbc.co.uk/programmes /p00ldwyp. Tim Harford also explores Jarrett's work and themes related to creativity under constraints in a 2019 *Cautionary Tales* episode—Tim Harford, "Bowie, Jazz, and the Unplayable Piaon," December 20, 2019, in *Cautionary Tales*, episode 7, produced by Ryan Dilley and Marilyn Rust, podcast, https://timharford.com/2019/12/cautionary-tales-ep-7-bowie-jazz-and-the -unplayable-piano/—and also a more in-depth treatment in Tim Harford, *Messy: How to Be Creative and Resilient in a Tidy-Minded World*, (London: Hachette UK, 2016).

22 "For a renowned perfectionist": Charles Waring, 'The Köln Concert': How Keith Jarrett Defied the Odds to Record His Masterpiece, uDiscover Music, January 24, 2022, accessed July 4, 2022, https: //www.udiscovermusic.com/stories/koln-concert-keith-jarrett/.

23 According to Eicher: Ibid.

23 "Inwardly [he was] very much": Gambaccini, "Keith Jarrett."

23 Toni Morrison: Toni Morrison, "Toni Morrison, the Art of Fiction No. 134," *The Paris Review*, 128 (Fall 1993), 1–33.

24 Not for nothing: Bill Janovitz, "The 40th Anniversary of Keith Jarrett's Legendary 'Köln Concert," *Observer*, March 12, 2015, accessed July 4, 2022, https://observer.com/2015/03/the-40th -anniversary-of-keith-jarretts-legendary-koln-concert/.

25 Whenever we face a stressful situation: The formative text on stress appraisals is Richard S. Lazarus and Susan Folkman, *Stress, Appraisal, and Coping* (New York: Springer Publishing Company, 1984); Matthias Jerusalem and Ralf Schwarzer, "Self-Efficacy as a Resource Factor in Stress Appraisal Processes," in *Self-efficacy: Thought Control of Action*, ed. Ralf Schwarzer (London: Hemisphere Publishing, 1992), 195–213; and Adrian Hase et al., "The Relationship Between Challenge and Threat States and Performance: A Systematic Review," *Sport, Exercise, and Performance Psychology* 8, no. 2 (2019): 123.

26 a large-scale and long-term US study: Abiola Keller et al., "Does the Perception that Stress Affects Health Matter? The Association with Health and Mortality," *Health Psychology* 31, no. 5 (2012): 677, as described in Kelly McGonigal, *The Upside of Stress: Why Stress Is Good for You, and How to Get Good at It* (New York: Penguin, 2016).

26 What is now becoming clear: This draws on McGonigal, *The Upside of Stress.*

28 One of the most powerful examples: Adrian Harvey et al., "Threat and Challenge: Cognitive Appraisal and Stress Responses in Simulated Trauma Resuscitations," *Medical Education* 44, no. 6 (2010): 587–94.

29 Neuroscientist-turned-psychologist Ian Robertson: Ian Robertson, *The Stress Test: How Pressure Can Make You Stronger and Sharper* (London: Bloomsbury Publishing, 2017).

29 As he puts it: Ian H. Robertson, "The Stress Test: Can Stress Ever Be Beneficial," *Journal of the British Academy* 5 (2017): 172–73.

30 Research by teams: Alison Wood Brooks, "Get Excited: Reappraising Pre-Performance Anxiety as Excitement," *Journal of Experimental Psychology: General* 143, no. 3 (2014): 1144.

30 Surgeons click into having better focus: A number of these examples draw from McGonigal, *The Upside of Stress.*

31 "All I remember": Grammy.com, 2014, accessed July 4, 2022, https://www.grammy.com/interview /the-making-of-keith-jarretts-the-k-ln-concert.

31 According to stress psychologist: Draws from an excellent profile in McGonigal, *The Upside of Stress.*

31 Such stress reappraisal mechanisms: "Getting butterflies to fly in formation" is from the title of a widely cited 1999 article in *The Sports Psychologist:* Sheldon Hanton and Graham Jones, "The Acquisition and Development of Cognitive Skills and Strategies: I. Making the Butterflies Fly in Formation," *The Sport Psychologist* 13, no. 1 (1999): 1–21.

32 "What happened with this piano": Keith Jarrett, produced by Mike Dibb, *The Art of Improvisation,* BBC, November 12, 2021, accessed July 14, 2022, https://www.bbc.co.uk/iplayer/episode /m0011f4y/keith-jarrett-the-art-of-improvisation.

32 "So, when I find": Janovitz, 2015.

33 Again in his words: Janovitz, 2015; and David Shenk, "Keith Jarrett, Part II: The Q&A," *The Atlantic,* October 13, 2009, accessed July 4, 2022, https://www.theatlantic.com/national/archive /2009/10/keith-jarrett-part-ii-the-q-a/27775/.

33 This is the space: NPR, Keith Jarrett and improvisation, 2017, accessed July 14, 2022, https://cpa .ds.npr.org/kuvo/audio/2017/04/jazz_film_jarrett_web.mp3.

33 From this click moment: Richard Lehnert, Meg Seaker, and Merridee Shaw, "Dancing on the Edge: Keith Jarrett on Music & Art," Stereophile.com, October 1, 1981, accessed July 4, 2022, https:// www.stereophile.com/content/dancing-edge-keith-jarrett-music-art.

34 Such a mindset was described: Carol Dweck, *Mindset: Changing the Way You Think to Fulfil Your Potential,* updated edition (London: Hachette UK, 2017).

37 A study of Australian students: Jacob J. Keech et al., "The Influence of University Students' Stress Mindsets on Health and Performance Outcomes," *Annals of Behavioral Medicine* 52, no. 12 (2018): 1046–59.

38 We can get a longer-term: Lew Hardy et al., "Great British Medallists: Psychosocial Biographies of Super-Elite and Elite Athletes from Olympic Sports," http://ipep.bangor.ac.uk/medalists _research.php.en, *Progress in Brain Research* 232 (2017): 1–119; and Arne Güllich et al., "Developmental Biographies of Olympic Super-Elite and Elite Athletes: A Multidisciplinary Pattern Recognition Analysis," *Journal of Expertise* 2, no. 1 (2019): 23–46.

40 The world record: Connie Suggitt, "56-year-old Free Diver Holds Breath for Almost 25 Minutes Breaking Record," May 12, 2021, accessed July 4, 2022, https://www.guinnessworldrecords.com /news/2021/5/freediver-holds-breath-for-almost-25-minutes-breaking-record-660285>; and freediving search results, accessed July 4, 2022, https://www.guinnessworldrecords.com/search ?term=freediving.

40 Their remarkable achievements: Per Fredrik Scholander, "The Master Switch of Life," *Scientific American* 209, no. 6 (1963): 92–107, https://www.scientificamerican.com/article/the-master-switch-of-life/.

41 *"During visualisation":* Harald Mathä, "An Interview with Katarina Linczenyiova, Freediver," Mares Scuba Diving Blog, June 10, 2016, accessed July 4, 2022, https://blog.mares.com/an-interview-with -katarina-linczenyiova-freediver-680.html.

41 *There is a growing body:* Teri Saunders et al., "The Effect of Stress Inoculation Training on Anxiety and Performance," *Journal of Occupational Health Psychology* 1, no. 2 (1996): 170.

42 *Recent research:* Sergio M. Pellis and Vivien C. Pellis, "What Is Play Fighting and What Is It Good For?" *Learning and Behavior* 45, no. 4 (2017): 355–66.

42 *This makes me think back:* These paragraphs extract and draw from a blog I first wrote for the Institute of Development Studies in 2017: Ben Ramalingam, "Powerful War Child Film Points Toward a New Kind of Ethos," December 7, 2017, accessed July 4, 2022, https://www.ids.ac.uk /opinions/powerful-war-child-film-points-towards-a-new-kind-of-aid-ethos/.

44 *His reply is exemplary:* Quoted in Hendrie Weisinger and J. P. Pawliw-Fry, *How to Perform Under Pressure: The Science of Doing Your Best When It Matters Most* (London: Hachette UK, 2015).

2. ORIGINALITY UNDER PRESSURE

45 *Originality Under Pressure:* The information on Red Flag exercises has numerous sources, including the Nellis Air Force Base public information Fact Sheets, "414th Combat Training Squadron 'Red Flag,'" July 6, 2012, accessed July 4, 2022, https://web.archive.org/web/20150918180334/http://www.nellis .af.mil/library/factsheets/factsheet.asp?id=19160; and Walter J. Boyne, "Red Flag," *Air Force Magazine*, November 2000, 47; John A. Tirpak, "Red Flag for the Future" *Air Force Magazine*, September 2014.

46 *"the best real-world training":* Royal Air Force, "RAF Flies the Flag on US Exercise," February 14, 2020, accessed July 14, 2022, https://www.raf.mod.uk/news/articles/raf-flies-the-flag-on-us-exercise/.

46 *More than thirty years:* Chelsey Sullenberger, curriculum vitae, June 19, 2019, accessed July 4, 2022, https://www.congress.gov/116/meeting/house/109642/witnesses/HHRG-116-PW05-Bio -SullenbergerC-20190619.pdf.

46 *"a moment during impending catastrophe":* Manjul Tripathi et al., "Sully, Simulation, and Neurosurgery," *World Neurosurgery* 118 (2018): 400–401.

47 *Prior to that fateful morning:* This section draws from numerous sources, including Sullenberger's autobiography and website. Chesley B. Sullenberger and Jeffrey Zaslow, *Highest Duty: My Search for What Really Matters* (New York: William Morrow, 2009); Chesley Sullenberger, n.d. "Navigating Crisis Beyond the Cockpit," accessed July 4, 2022, http://www.sullysullenberger.com/navigating -crisis-beyond-the-cockpit; and Kevin Kruse, "Stick the Landing: An Interview with Sully Sullenberger," July 21, 2017, accessed July 4, 2022, https://www.forbes.com/sites/kevinkruse/2017/07 /21/stick-the-landing-an-interview-with-sully-sullenberger/?sh=298861557106.

48 *He described his time:* Carl Von Wodkte, "Sully Speaks Out," September 7, 2016, accessed July 4, 2022, https://www.historynet.com/sully-speaks-out.

48 *Like the US Navy's:* Michael E. Weaver, "Missed Opportunities before Top Gun and Red Flag," *Air Power History* 60, no. 4 (2013): 18–31.

49 *In a roundabout way:* Lynne Martin, Jeannie Davison, Judith Orasanu, and Chesley Sullenberger, *Identifying Error Inducing Contexts in Aviation* (Warrendale, PA: SAE, Technical Paper No. 1999– 1–5540, 1999).

50 *By creating your list:* Joy Paul Guilford, "The Structure of Intellect," *Psychological Bulletin* 53, no. 4 (1956): 267; P. R. Christensen, J. P. Guilford, P. R. Merrifield, and R. C. Wilson, *Alternative*

Uses Test (Orange, CA: Sheridan Supply, 1960); and R. J. Sternberg and E. L. Grigorenko, "Guilford's Structure of Intellect Model and Model of Creativity: Contributions and Limitations," *Creativity Research Journal* 13, no. 3–4 (2001): 309–16.

51 **The reason for this shift**: Joy Paul Guilford, "Creativity," *American Psychologist* 5, no. 9 (1950): 444–54; Joy Paul Guilford, "Measurement and Creativity," *Theory into Practice* 5, no. 4 (1966): 185–89; Joy P. Guilford, "Creativity: Yesterday, Today and Tomorrow," *The Journal of Creative Behavior* no. 1 (1967): 3–14; and J. P. Guilford, "Characteristics of Creativity," Springield, IL: Illinois State Office of the Superintendent of Public Instruction, Gifted Children Section, 1973).

52 **When initially tasked**: Draws from the account provided in Fredrik Härén, *The Idea Book* (Stockholm: Interesting Books, 2004). This builds on the account in J. P. Guilford, *The Nature of Human Intelligence* (New York: McGraw-Hill, 1967).

53 **Convergent thinking *is the process***: J. P. Guilford, "Three Faces of Intellect," *The American Psychologist* 14 (1959): 469–79; and Harry D. Krop, Cecilia E. Alegre, and Carl D. Williams, "Effect of Induced Stress on Convergent and Divergent Thinking," *Psychological Reports* 24, no. 3 (1969): 895–98.

54 **If you want to assess**: Joy P. Guilford, "Creative Abilities in the Arts," *Psychological Review* 64, no. 2 (1957): 110; J. P. Guilford, "Can Creativity Be Developed?" *Art Education* 11, no. 6 (1958): 3–18; E. Paul Torrance, Carl H. Rush Jr., Hugh B. Kohn, and Joseph M. Doughty, *Factors in Fighter-Interceptor Pilot Combat Effectiveness* (Lackland Air Force Base, TX: Air Force Personnel and Training Research Center, 1957); and R. C. Wilson, J. P. Guilford, P. R. Christensen, and D. J. Lewis, "A Factor-Analytic Study of Creative-Thinking Abilities," *Psychometrika* 19, no. 4 (1954): 297–311.

54 **Research led by Guilford's colleagues**: Torrance et al., *Factors in Fighter-Interceptor Pilot Combat Effectiveness*.

56 **One of the most famous**: Draws from Pierluigi Serraino, *The Creative Architect: Inside the Great Midcentury Personality Study* (New York: Monacelli Press, 2016); and https://www.fastcompany.com/3060634/the-long-lost-study-that-tried-to-quantify-creative-personalities.

56 **The Alternative Uses Test**: Joy Paul Guilford, *Way Beyond the IQ* (Buffalo, NY: Creative Education Foundation, 1977).

57 **After the war, Guilford**: Guilford, "Creativity."

58 **In 1962**: Numerous sources including Elizabeth Schechter, *Self-Consciousness and 'Split' Brains: The Minds' I* (Oxford: Oxford University Press, 2018); and https://www.psychologytoday.com/gb/blog/the-theory-cognitive-modes/201404/sperry-jenkins-left-brain-right-brain.

59 **A series of famous experiments**: Michael S. Gazzaniga, "The Split-Brain: Rooting Consciousness in Biology," *Proceedings of the National Academy of Sciences* 111, no. 51 (2014): 18093–94; and Michael S. Gazzaniga, *Tales from Both Sides of the Brain: A Life in Neuroscience* (New York: Ecco/HarperCollins Publishers, 2015).

61 **As she puts it**: "Why Does It Work?—Drawing on the Right Side of the Brain," n.d., accessed July 4, 2022, https://www.drawright.com/theory, draws on Sperry's research to make this claim.

61 **Another bestselling creativity**: Julia Cameron, *The Artist's Way: A Spiritual Path to Higher Creativity* (Los Angeles: Tarcher, 1992), 11.

61 **Instead it works more like**: David Wolman, "The Split Brain: A Tale of Two Halves, *Nature* 483 (2012): 260–63, https://www.nature.com/articles/483260a.

62 In one of my roles: "Neuromyth 6," n.d., accessed July 4, 2022, https://www.oecd.org/education/ceri/neuromyth6.htm.

62 Research undertaken in 2012: Sanne Dekker et al., "Neuromyths in Education: Prevalence and Predictors of Misconceptions Among Teachers," *Frontiers in Psychology* (2012): 429.

63 Researchers at Harvard University: Roger E. Beaty et al., "Robust Prediction of Individual Creative Ability from Brain Functional Connectivity," *Proceedings of the National Academy of Sciences* 115, no. 5 (2018): 1087–92.

65 [This] would initially require: Alison Koontz, "The Circuitry of Creativity: How Our Brains Innovate Thinking," March 12, 2019, accessed July 4, 2022, https://caltechletters.org/science/what-is-creativity.

66 As a nice coda: Beaty et al., "Robust Prediction," 1087–92.

67 Longitudinal research undertaken in China: Jiangzhou Sun et al., "Training Your Brain to Be More Creative: Brain Functional and Structural Changes Induced By Divergent Thinking Training," *Human Brain Mapping* 37, no. 10 (2016): 3375–87.

69 In a world-famous TED Talk: Ken Robinson, "Do Schools Kill Creativity?" February 2006, TED Talk, video, 19:12, https://www.ted.com/talks/sir_ken_robinson_do_schools_kill_creativity.

70 It is more that: George Land and Beth Jarman, *Breakpoint and Beyond: Mastering the Future—Today* (New York: HarperCollins, 1993); George Land, "The Failure of Success," February 16, 2011, Tedx Talk, video, 13:06, https://www.youtube.com/watch?v=ZfKMq-rYtnc.

70 In schools, the methods: Mary Jo Puckett Cliatt, Jean M. Shaw, and Jeanne M. Sherwood, "Effects of Training on the Divergent-Thinking Abilities of Kindergarten Children," *Child Development* 51 (1980): 1061–64. A more recent study came to similar conclusions: Robyn E. Charles and Mark A. Runco, "Developmental Trends in the Evaluative and Divergent Thinking of Children," *Creativity Research Journal* 13, no. 3–4 (2001): 417–37.

71 Both teenagers and adults participating: Mathieu Hainselin, Alexandre Aubry, and Béatrice Bourdin, "Improving Teenagers' Divergent Thinking with Improvisational Theater," *Frontiers in Psychology* 9, no. 1759 (2018): 1–9; Diana Schwenke et al., "Improv to Improve: The Impact of Improvisational Theater on Creativity, Acceptance, and Psychological Well-Being," *Journal of Creativity in Mental Health* 16, no. 1 (2021): 31–48; and Clay Drinko, "Improv Boosts Creativity and Psychological Well-Being," May 18, 2020, accessed July 4, 2022, https://www.psychologytoday.com/gb/blog/play-your-way-sane/202005/improv-boosts-creativity-and-psychological-well-being.

71 They were then given: Benjamin Brooks et al., "New Human Capabilities in Emergency and Crisis Management: From Non-Technical Skills to Creativity," *The Australian Journal of Emergency Management* 34, no. 4 (2019): 23–30.

71 In a very different: Markus Baer and Greg R. Oldham, "The Curvilinear Relation Between Experienced Creative Time Pressure and Creativity: Moderating Effects of Openness to Experience and Support for Creativity," *Journal of Applied Psychology* 91, no. 4 (2006): 963.

72 Ken Robinson wrote a treatise: Ken Robinson and Lou Aronica, *Creative Schools: Revolutionizing Education from the Ground Up* (London: Penguin UK, 2015).

73 "self-defeating cul-de-sac": Tristram Hunt, "Creative Schools Review: We Need to Call Time on Exam-Factory Education," April 23, 2015, accessed July 4, 2022, https://www.theguardian.com/books/2015/apr/23/creative-schools-revolutionising-education-from-the-ground-up-ken-robinson-lou-aronica-review.

74 *Remarkably, the entirety:* University of Oxford, "Chimps Show Much Greater Genetic Diversity Than Humans," 2012, accessed July 4, 2022, https://www.ox.ac.uk/news/2012–03–02-chimps -show-much-greater-genetic-diversity-humans.

74 *Archaeologists have linked:* Curtis W. Marean, "When the Sea Saved Humanity," *Scientific American* 303, no. 2 (2010): 54–61.

75 *The prehistoric inhabitants of Blombos:* Janne-Beate Buanes Duke, "He Played Here as a Child. Then He Became an Archaeologist and Found a Now Famous Cave That Answers Questions of Our Past," January 13, 2020, accessed July 4, 2022, https://partner.sciencenorway.no/archaeology -humanities-technology/he-played-here-as-a-child-then-he-became-an-archaeologist-and -found-a-now-famous-cave-that-answers-questions-of-our-past/1617773.

75 *In 2018, a dig led:* Christopher S. Henshilwood et al., "An Abstract Drawing from the 73,000-Year-Old Levels at Blombos Cave, South Africa," *Nature* 562, no. 7725 (2018): 115–18; and *The Conversation*, "*South Africa's Blombos Cave Is Home to the Earliest Drawing by a Human,*" September 12, 2018, accessed July 4, 2022, https://theconversation.com/south-africas-blombos -cave-is-home-to-the-earliest-drawing-by-a-human-103017.

76 *"more salient, memorable":* Kristian Tylén et al., "The Evolution of Early Symbolic Behavior in Homo sapiens," *Proceedings of the National Academy of Sciences* 117, no. 9 (2020): 4578–84.

77 *As one of Henshilwood's:* Guy Gugliotta, "The Great Human Migration," July 2008, accessed July 4, 2022, https://www.smithsonianmag.com/history/the-great-human-migration-13561.

77 *Instead they used what the archaeologists:* Francesco d'Errico et al., "Identifying Early Modern Human Ecological Niche Expansions and Associated Cultural Dynamics in the South African Middle Stone Age," *Proceedings of the National Academy of Sciences* 114, no. 30 (2017): 7869–76.

77 *In one remarkable set of experiments:* E. Mellet et al., "Neuroimaging Supports the Representational Nature of the Earliest Human Engraving," *Royal Society Open Science* 6, no. 7 (2019): 190086, http://dx.doi.org/10.1098/rsos.190086.

78 *Researchers have concluded that:* Lyn Wadley, "Recognizing Complex Cognition through Innovative Technology in Stone Age and Palaeolithic Sites," *Cambridge Archaeological Journal* 23, no. 2 (2013): 163–83.

78 *The latest cross-continental genetic analysis:* Teresa Rito et al., "A Dispersal of Homo sapiens from Southern to Eastern Africa Immediately Preceded the Out-of-Africa Migration," *Scientific Reports* 9, no. 1 (2019): 1–10.

79 *To give the final word:* Guilford, "Creativity: Yesterday, Today and Tomorrow," 11.

3. THE STRENGTH OF PURPOSE

81 *The Strength of Purpose:* Numerous sources for the story of the Royal Victoria Belfast including: R. J. Barr and R. A. B. Mollan, "The Orthopaedic Consequences of Civil Disturbance in Northern Ireland," *The Journal of Bone and Joint Surgery* 71-B, no. 5 (November 1989): 739–44, https://doi.org /10.1302/0301–620X.71B5.2584241; "Bloody Friday: How the Troubles Inspired Belfast's Medical Pioneers," July 20, 2012, accessed July 14, 2022, https://www.bbc.com/news/uk-northern-ireland -18886867; Dermot P. Byrnes, "The Belfast Experience," in *Mass Casualties: A Lessons Learned Approach (Accidents, Civil Unrest, Natural Disasters, Terrorism)*, edited by R. Adams Cowley (Baltimore: U.S. Department of Transportation, 1983), 83–94; Francis X. Clines, "Ulster Doctors Learn to Deal with the Victims of Violence," *The New York Times*, August 18, 1987, https://www.nytimes.com

/1987/08/18/world/ulster-doctors-learn-to-deal-with-the-victims-of-violence.html; Ruth Coon, "How Northern Ireland's Doctors and Nurses Coped with the Troubles," *Brainstorm*, Raidió Teil-ifís Éireann, updated April 3, 2019; Richard Clarke, *The Royal Victoria Hospital, Belfast: A History 1797–1997* (Belfast: Blackstaff Press, 1997); Department of Foreign Affairs, *Post-Mortem Examinations Carried Out at Altnagelvin Hospital, Derry, on the Deceased Victims of the "Bloody Sunday" Shootings of 30 January, 1972* (Dublin: National Archives, Ireland, 1972, 2003/17/335), https://cain.ulster.ac.uk/nai/1972/nai_DFA-2003–17–335_1972–01–31_a.pdf; Marie-Therese Fay and Marie Smith, *Personal Accounts from Northern Ireland's Troubles: Public Conflict, Private Loss* (London: Pluto Press, 2000); Peter Froggatt, "Medicine in Ulster in Relation to the Great Famine and 'The Troubles,'" *British Medical Journal* 319, no. 7225 (December 1999): 1636, https://www.jstor.org/stable/25186692; Alf McCreary, "The Human Story of a Great Hospital," *Belfast Telegraph*, March 6, 1972; Alf McCreary, "The Human Story of a Great Hospital—Part Two," *Belfast Telegraph*, March 7, 1972; Phillip McGarry, "The Fortunes of the Legal and Medical Professions During the 'Troubles,'" *The Ulster Medical Journal* 84, no. 2 (October 2015): 119–23, https://www.ncbi.nlm.nih.gov/pmc/articles/PMC4488917/; James McKenna, Farhat Manzoor, and Greta Jones, *Candles in the Dark: Medical Ethical Issues in Northern Ireland during the Troubles* (London: Nuffield Trust, 2009). Farhat Manzoor, Greta Jones, and James McKenna, "'How Could These People Do This Sort of Stuff and Then We Have to Look After Them?' Ethical Dilemmas of Nursing in the Northern Ireland Conflict," *Oral History* 35, no. 2 (Fall 2007): 36–44, https://www.jstor.org/stable/40179944; Kate O'Hanlon, *Sister Kate: Nursing through the Troubles* (Belfast: Blackstaff Press, 2008); W. H. Rutherford, "Experience in the Accident and Emergency Department of the Royal Victoria Hospital with Patients from Civil Disturbances in Belfast 1969–1972, with a Review of Disasters in the United Kingdom 1951–1971," *Injury* 4, no. 3 (February 1973): 189–99, https://doi.org/10.1016/0020–1383(73)90038–7; W. H. Rutherford, "Surgery of Violence: II. Disaster Procedures," *British Medical Journal* 1, no. 5955 (February 1975): 443–45, https://doi.org/10.1136/bmj.1.5955.443; "The Thankless Task of Nursing the Troubles," *The Irish Times*, September 20, 2008, https://www.irishtimes.com/news/the-thankless-task-of-nursing-the-troubles-1.939955; and John Williams, "Casualties of Violence in Northern Ireland," *International Journal of Trauma Nursing* 3, no. 3 (July–September 1997): 78–82, https://doi.org/10.1016/S1075–4210(97)90033-X.

82 *One day somebody came:* O'Hanlon, *Sister Kate*.

83 *According to a BBC report:* "Bloody Friday," BBC News.

84 *Harvard historian Nancy Koehn:* Nancy Koehn, *Forged in Crisis: The Power of Courageous Leadership in Turbulent Times* (New York: Simon and Schuster, 2017); and Nancy Koehn, "Real Leaders Are Forged in Crisis," *Harvard Business Review* 3 (2020): 1–6.

84 *"This is the realization":* John Laidler, "People Become Leaders by Responding Effectively to Challenges, Author Says," *Harvard Gazette*, November 2, 2017, accessed July 4, 2022, https://news.harvard.edu/gazette/story/2017/11/people-become-leaders-by-responding-effectively-to-challenges-author-says/.

84 *Koehn's in-depth investigation:* Koehn, *Forged in Crisis*.

85 *The scientific interest:* Viktor Emil Frankl, *Recollections: An Autobiography*, trans. Joseph Fabry and Judith Fabry (New York: Insight Books/Plenum Press, 1997); and Viktor E. Frankl, *Man's Search for Meaning* (New York: Simon and Schuster, 1963).

86 *Frankl's book:* Catdir.loc.gov. Publisher description for Library of Congress Control Number 2006287144, n.d., accessed July 4, 2022, http://catdir.loc.gov/catdir/enhancements/fy0628/2006287144-d.html.

86 Data from the previously: Patrick L. Hill and Nicholas A. Turiano, "Purpose in Life as a Predictor of Mortality Across Adulthood," *Psychological Science* 25, no. 7 (2014): 1482–86; Patrick L. Hill et al., "The Value of a Purposeful Life: Sense of Purpose Predicts Greater Income and Net Worth," *Journal of Research in Personality* 65 (2016): 38–42; Patrick L. Hill et al., "Sense of Purpose Predicts Daily Positive Events and Attenuates Their Influence on Positive Affect," *Emotion* 22, no. 3 (2022): 597–602; and Midus.wisc.edu, *MIDUS Newsletter: Purpose in Life*, n.d., accessed July 4, 2022, http://www.midus.wisc.edu/newsletter/Purpose.pdf.

86 a separate study: Aliya Alimujiang et al., "Association Between Life Purpose and Mortality among US Adults Older Than 50 Years," *JAMA Network Open* 2, no. 5 (2019): e194270–e194270p.

86 One of the most influential scholars: T. M. Amabile, "The Social Psychology of Creativity: A Componential Conceptualization," *Journal of Personality and Social Psychology* 45, no. 2 (1983): 357–76; T. M. Amabile, *Creativity in Context* (Boulder, CO: Westview, 1996); and T. M. Amabile et al., "Assessing the Work Environment for Creativity," *Academy of Management Journal* 29 (1996): 1154–84.

87 "production of ideas or solutions": Amabile, *Creativity in Context.*

87 She has found four distinct ways: Teresa M. Amabile, Constance N. Hadley, and Steven J. Kramer, "Creativity Under the Gun," *Harvard Business Review* 80 (2002): 52–63.

89 "a man who constantly looked": John Rutherford and Jonathan Marrow, "William Harford Rutherford," *British Medical Journal* 336, no. 7649 (April 2008): 897.

90 "you have to love everybody": John Adams, "Kate O'Hanlon 1930–2014," *Nursing Standard* 29, no. 2 (2014): 34.

91 Koehn, in her analyses: Koehn, *Forged in Crisis.*

91 For aid workers: Thomas Morley, "A Bad Boss Is More Stressful Than War, Aid Workers Say," news .trust.org, January 26, 2006, accessed July 4, 2022, https://news.trust.org/item/20060126000000 -s0npj.

92 One of the most iconic Upshifters: Amy Berish, "FDR and Polio," FDR Presidential Library and Museum, n.d., accessed July 4, 2022, https://www.fdrlibrary.org/polio.

92 "Those who today are fortunate": Franklin Delano Roosevelt, "Statement on the New National Foundation for Infantile Paralysis," The American Presidency Project, 1937, accessed July 4, 2022, https://www.presidency.ucsb.edu/documents/statement-the-new-national-foundation-for -infantile-paralysis.

93 The reality: Liz Jackson, "Opinion: We Are the Original Lifehackers," *New York Times*, May 30, 2018, accessed July 4, 2022, https://www.nytimes.com/2018/05/30/opinion/disability-design -lifehacks.html.

93 "operate skillfully and inventively": Tim Edensor, ed., *Geographies of Rhythm: Nature, Place, Mobilities and Bodies* (Farnham: Ashgate Publishing, 2012).

93 "[they] are often": Paul Miller, Sophia Parker, and Sarah Gillinson, *Disablism: How to Tackle the Last Prejudice* (London: Demos, 2004).

94 "Franklin's illness": "Franklin Delano Roosevelt Memorial," U.S. National Park Service, n.d., accessed July 4, 2022, https://www.nps.gov/places/000/franklin-delano-roosevelt-memorial.htm.

94 Research in Italy: Mihaly Csikszentmihalyi, *Flow: The Psychology of Optimal Experience* (New York: Harper and Row, 1990).

94 "When I became paraplegic": Patient named Lucio, quoted in Csikszentmihalyi, *Flow.*

94 In the disability lifehacker movement: Ashley Shew, "Let COVID-19 Expand Awareness of Disability Tech," *Nature* 581 no. 7806 (2020): 9–10; and Hung Jen Kuo et al., "Current Trends in Technology and Wellness for People with Disabilities: An Analysis of Benefit and Risk," in *Recent Advances in Technologies for Inclusive Well-Being*, ed. Anthony Lewis Brooks et al. (Cham, Switzerland: Springer, 2021), 353–71.

95 Architect Betsey Farber: Liz Jackson, "Opinion: We Are the Original Lifehackers."

95 "We notice things": "About Us,"oxouk.com, n.d., accessed July 4, 2022, https://www.oxouk.com/aboutus.

95 John Elias: Staff writer, "iPhone Birthed from Purchase of Touch Pioneers FingerWorks?" *Wired,* January 22, 2007, accessed July 4, 2022, https://www.wired.com/2007/01/iphone-birthed-/.

96 Africa Humanitarian Action: For more about African Humanitarian Action, visit their website: https://africahumanitarian.org.

98 A fascinating study: Dean A. Shepherd, Fouad Philippe Saade, and Joakim Wincent, "How to Circumvent Adversity? Refugee-Entrepreneurs' Resilience in the Face of Substantial and Persistent Adversity," *Journal of Business Venturing* 35, no. 4 (2020): 105940.

99 Interviews with angel investors: Charles Y. Murnieks et al., "Drawn to the Fire: The Role of Passion, Tenacity and Inspirational Leadership in Angel Investing," *Journal of Business Venturing* 31, no. 4 (2016): 468–84.

100 As Victor Frankl found: Frankl, *Man's Search for Meaning.*

4. THE POWER OF UPSHIFT

101 How is it possible: NASA, "Apollo 13," July 8, 2009, accessed July 4, 2022, https://www.nasa.gov/mission_pages/apollo/missions/apollo13.html; "Women Who Changed Science: Tu Youyou, Nobel Prize, n.d., accessed July 4, 2022, https://www.nobelprize.org/womenwhochangedscience/stories/tu-youyou; Zhang Jianfang, *A Detailed Chronological Record of Project 523 and the Discovery and Development of Qinghaosu (Artemisinin)* (Jupiter, FL: Strategic Book Publishing, 2013); Cui Weiyuan, "Ancient Chinese Anti-Fever Cure becomes Panacea for Malaria," *World Health Organization, Bulletin of the World Health Organization* 87, no. 10 (2009): 743; and Timothy C. Winegard, *The Mosquito: A Human History of Our Deadliest Predator* (Text Publishing, 2019).

102 It has become shorthand: Alex Davies, "Why 'Moon Shot' Has No Place in the 21st Century," *Wired,* July 15, 2019, https://www.wired.com/story/apollo-11-moonshot-21st-century/.

103 In the Apollo 13 movie: *Apollo 13*, directed by Ron Howard, featuring Tom Hanks, Kevin Bacon, Bill Paxton, and Ed Harris, released June 30, 1995, distributed by Universal Pictures.

104 But it never actually happened: Colin Burgess, ed., *Footprints in the Dust: The Epic Voyages of Apollo, 1969–1975* (Lincoln: University of Nebraska Press, 2010).

104 "If you spend your time": Kevin Fong, "50 Years On—How *Apollo 13*'s Near Disastrous Mission Is Relevant Today," *Guardian,* February 29, 2020, accessed July 4, 2022, https://www.theguardian.com/science/2020/feb/29/apollo-13-how-teamwork-and-tenacity-turned-disaster-into-triumph.

104 "Everybody started talking": Burgess, *Footprints in the Dust.*

104 "I never felt": Andrew Chaikin, "Apollo 13 Astronauts Share Surprises from their 'Successful Failure' Mission," CollectSPACE.com, n.d., accessed July 4, 2022, http://www.collectspace.com/news/news-041310a.html.

104 "Some years later": "Apollo Expeditions to the Moon: Chapter 13," History.nasa.gov, n.d. accessed July 4, 2022, https://history.nasa.gov/SP-350/ch-13-4.html.

104 Because their work: Tu Thanh Ha, "The Secret Military Project That Led to a Nobel Prize," *Globe and Mail*, accessed July 4, 2022, https://www.theglobeandmail.com/news/world/scientists -share-nobel-medicine-prize-for-work-to-fight-parasitic-diseases/article26648532/.

105 "Project 523 became": Zhang, *A Detailed Chronological Record*.

105 The US effort: Youyou Tu, "Artemisinin: A Gift from Traditional Chinese Medicine to the World," Nobel lecture, December 7, 2015, accessed July 14, 2022, https://www.nobelprize.org /uploads/2018/06/tu-lecture.pdf.

105 The original project document: Jia Chen-Fu, "The Secret Maoist Chinese Operation That Conquered Malaria—and Won a Nobel," *Conversation*, October 6, 2015, accessed July 4, 2022, https: //theconversation.com/the-secret-maoist-chinese-operation-that-conquered-malaria-and-won-a -nobel-48644.

106 "Shennong tasted": Youyou Tu, "The Nobel Prize in Physiology or Medicine 2015," Nobel-Prize.org, 2016, accessed July 4, 2022, https://www.nobelprize.org/prizes/medicine/2015/tu /biographical.

106 "We were given the situation": Olugbemisola Rhuday-Perkovich, *Above and Beyond: NASA's Journey to Tomorrow* (New York: Feiwel and Friends, 2018).

107 "We never were": A. Chaikin and G. Lunney, "Glynn S. Lunney Apollo 13 Narrative," NASA Johnson Space Center Oral History Project, n.d., accessed July 4, 2022, https://historycollection .jsc.nasa.gov/JSCHistoryPortal/history/oral_histories/LunneyGS/Apollo13.htm; Melanie Whiting, ed., "NASA Icons Showcase Lunar Leadership," NASA History, April 3, 2018, accessed July 4, 2022, https://www.nasa.gov/feature/nasa-icons-showcase-lunar-leadership; and N. Atkinson, "13 MORE Things That Saved Apollo 13, part 2: Simultaneous Presence of Kranz and Lunney at the Onset of the Rescue," *Universe Today*, 2015, accessed July 4, 2022, https://www.universetoday.com /119778/13-more-things-that-saved-apollo-13-part-2-simultaneous-presence-of-kranz-and-lunney -at-the-onset-of-the-rescue/.

108 "In the midst of this": T. Mattingly II, 2001, "Edited Oral History Transcript," NASA Johnson Space Center Oral History Project, 2001, accessed July 14, 2022 https://historycollection.jsc.nasa .gov/JSCHistoryPortal/history/oral_histories/MattinglyTK/MattinglyTK_11-6-01.htm.

108 There was a clear sense: Zhang, *A Detailed Chronological Record*.

109 They analyzed more than: Park Jong Yeon, "Artemisinin and the Nobel Prize in Physiology or Medicine 2015," *The Korean Journal of Pain* 32, no. 3 (2019): 145–146; Zheng Wei-Rong et al., "Tu Youyou Winning the Nobel Prize: Ethical Research on the Value and Safety of Traditional Chinese Medicine," *Bioethics* 34, no. 2 (2020): 166–171; and Wang Jigang et al., "Artemisinin, the Magic Drug Discovered from Traditional Chinese Medicine," *Engineering* 5, no. 1 (2019): 32–39.

110 "[Tu's] institute had a good start": Zhang, *A Detailed Chronological Record*.

110 "it was a question": Burgess, *Footprints in the Dust*.

111 "There was no room": T. Mattingly II, "Edited Oral History Transcript."

111 "[There was] a sense": Zhang, *A Detailed Chronological Record*.

112 "the work was the top priority": E. Andrew Balas, *Innovative Research in Life Sciences: Pathways to Scientific Impact, Public Health Improvement, and Economic Progress* (Hoboken, NJ: Wiley, 2018).

112 Yiqing recalled: Weiyuan, "Ancient Chinese Anti-Fever Cure," 743.

112 "very laborious and tedious job": Ushma S. Neill, "From Branch to Bedside: Youyou Tu Is Awarded the 2011 Lasker~DeBakey Clinical Medical Research Award for Discovering Artemisinin as a Treatment for Malaria," *The Journal of Clinical Investigation* 121, no. 10 (2011): 3768–73.

113 "Our desire [to] have": Youyou Tu, "The Nobel Prize in Physiology or Medicine 2015."

113 In 2018, 214 million: "The 'World Malaria Report 2019' at a Glance," World Health Organization, December 4, 2019, accessed July 4, 2022, https://www.who.int/news-room/feature-stories /detail/world-malaria-report-2019.

113 In absolute numbers: "Malaria," World Health Organization, 2021, accessed July 4, 2002, https: //www.who.int/data/gho/data/themes/malaria.

114 "the most important pharmaceutical intervention": "The Nobel Prize: Women Who Changed Science: Tu Youyou," Nobelprize.org, n.d., accessed July 14, 2022, https://www.nobelprize.org /womenwhochangedscience/stories/tu-youyou.

114 "It was a bit of a surprise": Chris Buckley, "Some Surprise, and Affirmation, in China After Tu Youyou Receives Nobel Prize," Sinosphere Blog, October 6, 2015, accessed July 4, 2022, https: //sinosphere.blogs.nytimes.com/2015/10/06/nobel-china-medicine-tu-youyou-prize/.

5. CHALLENGERS

122 "I realised": Colin Murphy, "Interview: Steve Collins on Famine Relief, January 14, 2009, accessed July 4, 2022, http://colinmurphy.ie/?p=256.

122 "In Sudan": Ibid.

124 "There was no data": "Disrupting Charity: How Social Business Can Eradicate Starvation," Ashoka Brasil, August 2, 2016, accessed July 4, 2022, https://www.ashoka.org/pt-br/story /disrupting-charity-how-social-business-can-eradicate-starvation.

124 In his subsequent publication: Steve Collins, "The Limit of Human Adaptation to Starvation," *Nature Medicine* 1, no. 8 (1995): 810–14.

125 Our brains use: Michael W. Richardson, "How Much Energy Does the Brain Use?" Brainfacts .org, February 1, 2019, accessed July 4, 2022, https://www.brainfacts.org/brain-anatomy-and -function/anatomy/2019/how-much-energy-does-the-brain-use-020119. Ferris Jabr, "Does Thinking Really Hard Burn More Calories?" *Scientific American*, accessed July 14, 2022, https://www .scientificamerican.com/article/thinking-hard-calories.

126 It turns out that almost half: Wendy Wood and David T. Neal, "The Habitual Consumer," *Journal of Consumer Psychology* 19, no. 4 (2009): 579–92.

126 "the typical human tendency": Richard P. Bagozzi and Kyu-Hyun Lee, "Consumer Resistance to, and Acceptance of, Innovations," in *ACR North American Advances NA—Advances in Consumer Research Volume* 26, eds. Eric J. Arnould and Linda M. Scott (Provo, UT: Association for Consumer Research, 1999), 218–25.

127 Gregory Berns, a neuroscientist: Gregory Berns, *Iconoclast: A Neuroscientist Reveals How to Think Differently* (Boston: Harvard Business Press, 2010).

127 Maggie Boden: Margaret A. Boden, *The Creative Mind: Myths and Mechanisms* (London: Routledge, 2004).

128 The journalist Tim Harford: Tim Harford, *Messy: How to Be Creative and Resilient in a Tidy-Minded World* (London: Hachette UK, 2016).

128 This was a remarkable study: Shaun Larcom, Ferdinand Rauch, and Tim Willems: "The Benefits of Forced Experimentation: Striking Evidence from the London Underground Network," *The Quarterly Journal of Economics* 132, no. 4 (2017): 2019–255.

129 According to Wendy Wood: Wendy Wood, *Good Habits, Bad Habits: The Science of Making Positive Changes That Stick* (New York: Pan Macmillan, 2019).

130 "sometimes the obstacle": Tim Harford, "The Doris Day Effect—When Obstacles Help Us," Timharford.com, June 14, 2019, accessed July 4, 2022, https://timharford.com/2019/06/the -doris-day-effect-when-obstacles-help-us.

130 Another example: Larcom et al., "The Benefits of Forced Experimentation," 2019–55.

131 Some valuable lessons: Gary A. Klein, *Sources of Power: How People Make Decisions* (Boston: MIT Press, 2017); and Gary Klein, *Seeing What Others Don't: The Remarkable Ways We Gain Insights* (New York: Public Affairs, 2013).

134 Lundgren had many problems: Jérôme Barthélemy, "The Experimental Roots of Revolutionary Vision," *MIT Sloan Management Review* 48, no. 1 (2006): 81.

134 "Why not take off the legs?": "The History of the Revolutionary IKEA Flatpacks," IKEA.com, n.d., accessed July 4, 2022, https://www.ikea.com/ph/en/this-is-ikea/about-us/the-story-of -ikea-flatpacks-puba710ccb0.

135 Harvard professor Clayton Christensen argued: Clayton M. Christensen, *The Innovator's Dilemma: When New Technologies Cause Great Firms to Fail* (Boston: Harvard Business Review Press, 1997).

136 "One of the greatest pains": Walter Bagehot, *Physics and Politics* (New York: Knopf: n.d.).

137 Watching the footage now: Amelia Butterly, "Thirty Years of Talking About Famine in Ethiopia— Why's Nothing Changed?" BBC News, November 11, 2015, accessed July 4, 2022, https://www .bbc.co.uk/news/newsbeat-34776109.

138 With his experiences of Sudan: Ben Ramalingam, "New Ideas Can Transform Aid Delivery," *Guardian*, February 22, 2011, accessed July 14, 2022, https://www.theguardian.com/global -development/poverty-matters/2011/feb/22/humanitarian-aid-innovation. Some of this material presented here draws from research undertaken as part a 2009 study I led on humanitarian innovation: Ben Ramalingam, Kim Scriven, and Conor Foley, *Innovations in International Humanitarian Action* (London: Overseas Development Institute, 2009).

139 They came in the form: A. Briend et al., "Ready-to-Use Therapeutic Food for Treatment of Marasmus," *Lancet* 353, no. 9166 (1999): 1767–68.

140 Armed with the knowledge: Steve Collins, "Changing the Way We Address Severe Malnutrition During Famine," *Lancet* 358, no. 9280 (2001): 498–501.

141 "In 2000, I went to Ethiopia": Colinmurphy.ie, 2009.

142 The international president: "Niger: Peanut-Based Wonder-Food Needs Wider Use, Relief Web, September 4, 2006, accessed July 14, 2022, https://reliefweb.int/report/niger/niger-peanut -based-wonder-food-needs-wider-use.

144 On March 5, 1770: Mitch Kachun, *First Martyr of Liberty: Crispus Attucks in American Memory* (New York: Oxford University Press, 2017); John Adams, "Speech by John Adams at the Boston Massacre Trial," Bostonmassacre.net, n.d., accessed July 4, 2022, http://www.bostonmassacre.net /trial/acct-adams3.htm; and Marcus Rediker, *Outlaws of the Atlantic: Sailors, Pirates, and Motley Crews in the Age of Sail* (Boston: Beacon Press, 2015).

146 When they took over a ship: Simon Worrall, "Q&A: Were Modern Ideas—and the American Revolution—Born on Ships at Sea?" *National Geographic*, August 29, 2014, accessed July 4, 2022, https://www.nationalgeographic.com/culture/article/140831-pirates-horatio-nelson-samuel-adams-royal-navy-somalia-ngbooktalk.

147 "The crews organised themselves": Peter Linebaugh and Marcus Rediker, "A Motley Crew in the American Revolution," Versobooks.com, July 3, 2016, accessed July 4, 2022, https://www.versobooks.com/blogs/2749-a-motley-crew-in-the-american-revolution; and John K. Alexander, *Samuel Adams: America's Revolutionary Politician* (Lanham, MD: Rowman and Littlefield, 2002).

148 "Joy lies in the fight": Mohandas Karamchand Gandhi, *My Religion* (Prabhat Prakashan, 2021).

6. CRAFTERS

149 Crafters: Overall sources: John Uri, "35 Years Ago: Remembering *Challenger* and Her Crew," NASA Johnson Space Center, January 28, 2021, accessed July 4, 2022, https://www.nasa.gov/feature/35-years-ago-remembering-challenger-and-her-crew; Colin Burgess, *Teacher in Space: Christa McAuliffe and the* Challenger *Legacy* (Lincoln: University of Nebraska Press, 2000); John F. Muratore, "NASA Johnson Space Center Oral History Project Tacit Knowledge Capture Project Edited Oral History Transcript," July 16, 2010, accessed July 14, 2022, https://historycollection.jsc.nasa.gov/JSCHistoryPortal/history/oral_histories/SSP/MuratoreJF_5–14–08.htm; *NASA Procurement in the Earth-Space Economy: Hearing Before the House Comm. on Science*, 104th Cong. (November 8, 1995); and Laurence Gonzales, "Lost Space: Outlaw Engineers Are Struggling to Save NASA with a New Generation of Secret Technology," *Rolling Stone*, April 6, 1995.

152 NASA management: Richard C. Cook, "Why I Blew the Whistle on NASA's O-Ring Woes," *Washington Post*, March 16, 1986, accessed July 14, 2022, https://www.washingtonpost.com/archive/opinions/1986/03/16/why-i-blew-the-whistle-on-nasas-o-ring-woes/d1711a71–7581–4c66-b037–5ed98a24583d/; and Joe Atkinson, "Engineer Who Opposed Challenger Launch Offers Personal Look at Tragedy," Nasa.gov, October 5, 2012, accessed July 14, 2022, https://www.nasa.gov/centers/langley/news/researchernews/rn_Colloquium1012.html.

153 "one computer would come in": Muratore, "NASA Johnson Space Center Oral History Project."

154 "on one side of": Gonzales, "Lost Space."

156 Scholars at: David Pye, *The Nature and Art of Workmanship* (Cambridge University Press, 1968).

156 "that productive, uncomfortable terrain": Daniel Coyle, *The Talent Code: Unlocking the Secret of Skill in Maths, Srt, Music, Sport, and Just About Everything Else* (London: Random House, 2009).

157 "What's the subject at hand?'": Muratore, "NASA Johnson Space Center Oral History Project."

161 Dr. Wendell Scott: Ryan Reft, "Charles and Ray Eames: How Wartime L.A. Shaped the Mid-Century Modern Aesthetic," KCET, September 1, 2016, accessed July 14, 2022, https://www.kcet.org/shows/lost-la/charles-and-ray-eames-how-wartime-l-a-shaped-the-mid-century-modern-aesthetic; and Eames.com, "Wartime," n.d., accessed July 4, 2022, https://eames.com/en/wartime; and https://www.eamesoffice.com/works/ww2; and Eamesoffice.com. "WWII," n.d., accessed July 4, 2022, https://www.eamesoffice.com/works/ww2.

162 "[The splint] was an": Eamesoffice.com. "Molded Plywood Leg Splint," n.d., accessed July 14, 2022, https://www.eamesoffice.com/the-work/molded-plywood-leg-splint/.

163 "artist-engineers": Leon Ransmeier, "Charles and Ray Eames and Their History of Plywood," https://archive.pinupmagazine.org, n.d., accessed July 4, 2022, https://pinupmagazine.org/articles/charles-ray-eames-history-of-plywood-with-herman-miller-by-leon-ransmeier.

163 *"creativity in the face of constraints"*: Charles Eames (1972), "Q&A Charles Eames in Design," (interview with Madame L'Amic for the Exhibition What Is Design?, held at the Louvre in 1969, later turned into a short film Design Q&A for Herman Miller; paper in Eames Archive.

164 *One of the most powerful*: Genrikh Saulovich Altshuller, *The Innovation Algorithm: TRIZ, Systematic Innovation and Technical Creativity* (Worcester, MA: Technical Innovation Center, 1999).

165 *Here are the things*: This draws on and adapts Nesta's Social Design Tools: Geoff Mulgan, "Design in Public and Social Innovation: What Works and What Could Work Better," NESTA January 2014, accessed July 4, 2022, https://media.nesta.org.uk/documents/design_in_public_and_social_innovation.pdf.

168 *Anyone who follows sports*: Simone Biles, *Courage to Soar (with Bonus Content): A Body in Motion, A Life in Balance* (Grand Rapids, MI: Zondervan, 2016); Alice Park "These Are All the Gymnastics Moves Named After Simone Biles," *Time*, July 26, 2021, accessed July 4, 2022, https://time.com/6083539/gymnastics-moves-named-after-simone-biles; Juliet Macur, "Simone Biles Dials Up the Difficulty, 'Because I Can,'" Nytimes.com. May 24, 2021, accessed July 4, 2022 https://www.nytimes.com/2021/05/24/sports/olympics/simone-biles-yurchenko-double-pike.html; and Simone Biles, "Simone Biles Builds Gymnastics Routine Out of Bricks!," Lego.com, n.d., accessed July 4, 2022, https://www.lego.com/en-us/kids/videos/lego/llsomeengymnastics-b1a06dc8d6294e56bca608a3e6a76149.

170 *As one analyst described it*: Gonzales, "Lost Space."

172 *"Our people aren't"*: Muratore, "NASA Johnson Space Center Oral History Project."

174 *"The Pirates' motto"*: Loizos Haracleous et al., "How a Group of NASA Renegades Transformed Mission Control," *MIT Sloan Management Review*, April 5, 2019, accessed July 4, 2022, https://sloanreview.mit.edu/article/how-a-group-of-nasa-renegades-transformed-mission-control/.

174 *"We have to change "*: *NASA Procurement in the Earth-Space Economy.*

7. COMBINERS

175 *Combiners*: H. Cai et al., "A Draft Genome Assembly of the Solar-Powered Sea Slug *Elysia chlorotica*," *Scientific Data* 6, no. 1 (2019): 1–13; and Lin Edwards, "Green Sea Slug Makes Chlorophyll Like a Plant," Phys.org, January 12, 2010, accessed July 4, 2022, https://phys.org/news/2010–01-green-sea-slug-chlorophyll.html.

178 *In a charming TED animation*: Catherine Mohr, "How I Became Part Sea Urchin," TED, video, 6:07, 2018, accessed July 4, 2022, https://www.ted.com/talks/catherine_mohr_how_i_became_part_sea_urchin.

183 *The ancient Egyptian and Chinese*: Kenneth Caiman, "The Arrow or the Caduceus as the Symbol of the Doctor," *The Lancet* 362, no. 9377 (2003): 84.

184 *The unique value of Combiners*: John Bessant and Anna Trifilova, "Developing Absorptive Capacity for Recombinant Innovation," *Business Process Management Journal* 23, no. 6 (2017): 1094–1107; and John Bessant, "Bridging Different Worlds—The Power of Recombinant Innovation," Blog.hypeinnovation.com, 2016, accessed July 4, 2022, https://blog.hypeinnovation.com/the-power-of-recombinant-innovation.

186 *"Over the years"*: Jonathan Noble, "Hamilton: DAS Innovation Came from "Breaking" Engineers," motorsport.com, March 5, 2020, accessed July 4, 2022, https://au.motorsport.com/f1/news/hamilton-mercedes-das-breaking-engineers/4719127/.

187 *"You can always improve"*: "Lewis Hamilton Reveals He Changes Driving Style 'Every Year' in Relentless Bid to Improve," Formula1.com., accessed July 4, 2022, https://www.formula1.com/en/latest/article.hamilton-reveals-he-changes-driving-style-every-year-in-relentless-bid-to.6BkNo1SmhGslMYA87VC37o.html.

188 *"Diamonds are created"*: J. Wells, "Lewis Hamilton Is Not a Racing Driver," *Gentleman's Journal*, May 2, 2020, accessed July 4, 2022, https://www.thegentlemansjournal.com/article/lewis-hamilton-is-not-a-racing-driver-interview-formula-one-met-gala/.

189 *War and peas*: Nicolas Appert, *The Art of Preserving All Kinds of Animal and Vegetable Substances for Several Years: A Work Published by the Order of the French Minister of the Interior, on the Report of the Board of Arts and Manufacturers*, vol. 1 (London: Black, Parry, and Kingsbury, 1812); Lindsay Evans, "Nicolas Appert," in *Science and Its Times: Understanding the Social Significance of Scientific Discovery* Volume 5, eds. Neil Schjlager and Josh Lauer (Farmington Hills, MI: Gale Group, 2001); Rebeca Garcia and Jean Adrian, "Nicolas Appert: Inventor and Manufacturer," *Food Reviews International* 25, no. 2 (2009): 115–25; and Susan Featherstone, "A Review of Development in and Challenges of Thermal Processing Over the Past 200 Years—A Tribute to Nicolas Appert," *Food Research International* 47, no. 2 (2012): 156–60.

190 *"Having spent my time"*: Appert, *The Art of Preserving*.

195 *Beloved Lina*: Mitch Waldrop, "Inside Einstein's Love Affair with 'Lina'—His Cherished Violin. *National Geographic*, February 3, 2017, accessed July 4, 2022, https://www.nationalgeographic.com/adventure/article/einstein-genius-violin-music-physics-science; "The Symphony of Science," NobelPrize.org. Nobel Prize Outreach AB 2022. 2, March 2019, accessed July 4, 2022, https://www.nobelprize.org/symphony-of-science/; "You Don't Have to Be Einstein to Play an Instrument, Although He Did," Music Nation, September 18, 2017, accessed July 4, 2022, https://musication.nyc/einstien-played-an-instrument; Justin Chandler and Tahiat Mahboob, "Albert Einstein: 10 Things You Might Not Know About His Love for Music," CBC Music, March 14, 2017, accessed July 4, 2022, https://www.cbc.ca/music/read/albert-einstein-10-things-you-might-not-know-about-his-love-for-music-1.5073973; Arthur I. Miller, "A Genius Finds Inspiration in the Music of Another," UCL News, January 31, 2006, accessed July 4, 2020, https://www.ucl.ac.uk/news/2006/jan/genius-finds-inspiration-music-another; and Walter Sullivan, "The Einstein Papers: Childhood Showed a Gift for the Abstract," *New York Times*, March 27, 1972, accessed July 4, 2022, https://www.nytimes.com/1972/03/27/archives/the-einstein-papers-childhood-showed-a-gift-for-the-abstract-the.html.

8. CONNECTORS

203 *For Maia Majumder*: Maimuna S. Majumder, "Coronavirus Researchers Are Dismantling Science's Ivory Tower—One Study at a Time," *Wired*, June 18, 2020, accessed July 4, 2022, www.wired.com/story/covid-19-studies-dismantle-science-ivory-tower.

206 *If you are fortunate enough*: Edward Bishop Smith, Tanya Menon, and Leigh Thompson, "Status Differences in the Cognitive Activation of Social Networks," *Organization Science* 23, no. 1 (2012): 67–82.

207 *A recent study looked*: B. Kovács et al., "Social Networks and Loneliness during the COVID-19 Pandemic," *Socius* 7 (2021): 1–16; Marissa King and Balázs Kovács, "Research: We're Losing

Touch with Our Networks," *Harvard Business Review*, February 12, 2021, accessed July 4, 2022, https://hbr.org/2021/02/research-were-losing-touch-with-our-networks.

208 One answer comes: L. Barabási, *The Formula: The Universal Laws of Success* (New York: Little, Brown, 2018); Manuel Castells, *Communication Power*, 2nd ed. (Oxford: Oxford University Press, 2013); and Duncan J. Watts and S. Strogatz, *Six Degrees: Science in a Connected Age* (New York: Norton, 2013).

210 Analysis of some nine thousand: Daniel M. Romero, Brian Uzzi, and Jon Kleinberg, "Social Networks Under Stress," *Proceedings of the 25th International Conference on World Wide Web* (2016).

211 the diagram below: Catherine T. Shea et al., "The Affective Antecedents of Cognitive Social Network Activation," *Social Networks* 43 (2015): 91–99.

213 Bridges, Bonds, and Links: David Hackett Fischer, *Paul Revere's Ride* (Oxford: Oxford University Press, 1994); Malcolm Gladwell, *The Tipping Point: How Little Things Can Make a Big Difference* (Boston: Little, Brown, 2006); Shin-Kap Han, "The Other Ride of Paul Revere: The Brokerage Role in the Making of the American Revolution," *Mobilization: An International Quarterly* 14, no. 2 (2009): 143–62; Brian Uzzi and Shannon Dunlap, "How to Build Your Network," *Harvard Business Review* 83, no. 12 (2005): 53, accessed July 4, 2022, https://hbr.org/2005/12/how-to-build-your-network; Kieran Healy, "Using Metadata to Find Paul Revere," Kieranhealy.org, June 9, 2013, accessed July 4, 2022, http://kieranhealy.org/blog/archives/2013/06/09/using-metadata-to-find-paul-revere; S. White, A. Enright and C. Brummitt, "Analyzing Social Networks of Colonial Boston Revolutionaries with the Wolfram Language," Blog.wolfram.com, June 29, 2017, accessed July 4, 2022, https://blog.wolfram.com/2017/06/29/analyzing-social-networks-of-colonial-boston-revolutionaries-with-the-wolfram-language; and Pauline Maier, "Making the Redcoats Look Silly," *New York Times*, April 17, 1994, accessed July 4, 2022, https://www.nytimes.com/1994/04/17/books/making-the-redcoats-look-silly.html.

217 "Local structures are already": William J. Clinton, foreword to *Joint Evaluation of the International Response to the Indian Ocean Tsunami: Synthesis Report*, by John Telford and John Cosgrave (London: Tsunami Evaluation Coalition, 2006).

217 The idea of social capital: Robert D. Putnam, "Bowling Alone: America's Declining Social Capital," in *Culture and Politics*, ed. Lane Crothers and Charles Lockhart (New York: Palgrave Macmillan, 2000), 223–234.

222 "When we arrived I noticed": Lyndall Stein, "The Power of the Crowd in the Internet Age," Fair Observer, March 1, 2017, accessed July 4, 2022, https://www.fairobserver.com/region/europe/public-protest-resist-trump-internet-culture-news-10882/.

9. CORROBORATORS

227 Corroborators: Draws from my reading of Sabrina Cohen-Hatton, *The Heat of the Moment: Life and Death Decision-Making from a Firefighter* (London: Transworld Digital, 2019).

231 In the 1980s: Richard H. Thaler and H. M. Shefrin, "An Economic Theory of Self-Control," *Journal of Political Economy* 89, no. 2 (1981): 392–406.

231 Another Nobel laureate, Daniel Kahneman: Daniel Kahneman, *Thinking, Fast and Slow* (New York: Macmillan, 2011).

234 Stress and trauma specialist: Elizabeth A. Stanley, *Widen the Window: Training Your Brain and Body to Thrive During Stress and Recover from Trauma* (New York: Penguin, 2019); Kelsey L. Larsen and Elizabeth A. Stanley, "Leaders' Windows of Tolerance for Affect Arousal—and Their Effects on Political Decision-making During COVID-19," *Frontiers in Psychology* 12 (2021): 749715; and Elizabeth A. Stanley, "War Duration and the Micro-Dynamics of Decision Making Under Stress," *Polity* 50, no. 2 (2018): 178–200.

235 *"The fact that Jarrett":* Anna Chesner, ed., *Trauma in the Creative and Embodied Therapies: When Words Are Not Enough* (Oxford and New York: Routledge, 2020).

238 *By age six: Florence Nightingale at Prayer—The Collected Works of Florence Nightingale*, accessed July 4, 2022, https://cwfn.uoguelph.ca/spirituality/florence-nightingale-at-prayer.

238 *Unusually named after:* Lynn McDonald, ed., *Collected Works of Florence Nightingale*, vol. 7, *Florence Nightingale's European Travels* (Waterloo, Canada: Wilfrid Laurier University Press, 2006); and Lynn McDonald, ed., *Collected Works of Florence Nightingale*, vol. 1, *Florence Nightingale: An Introduction to Her Life and Family* (Waterloo, Canada: Wilfrid Laurier University Press, 2001).

240 *Of particular note:* Marjie Bloy, Florence Nightingale (1820–1910), Victorianweb.org, n.d., accessed July 4, 2022, https://victorianweb.org/history/crimea/florrie.html; Lynn McDonald, ed., *Collected Works of Florence Nightingale*, vol. 14., *Florence Nightingale: The Crimean War* (Waterloo, Canada: Wilfrid Laurier University Press, 2010); and Cecil Woodham-Smith, *Lady-in-Chief: The Story of Florence Nightingale* (London: Methuen, 1953).

240 *In January and February:* Philip A. Mackowiak, "Florence Nightingale's Actual Cause of Death," OUPblog, August 13, 2015, accessed July 4, 2022, https://blog.oup.com/2015/08/florence-nightingale-syphilis-death.

241 *"useful for hard-pressed":* and other quotes from Ida Beatrice O'Malley, *Florence Nightingale, 1820–1856: A Study of Her Life Down to the End of the Crimean War* (London: Thornton Butterworth, 1931).

242 *"She wanted data for each regiment":* L. McDonald, *01. Florence Nightingale, Statistics and the Crimean War—The Collected Works of Florence Nightingale*, Cwfn.uoguelph.ca, 2017, accessed July 4, 2022, https://cwfn.uoguelph.ca/short-papers-excerpts/nightingale-statistics-and-the-crimean-war/; and Eileen Magnello, "Florence Nightingale: The Compassionate Statistician," *Plus Magazine from the Cambridge University Millennium Mathematics Project*, December 8, 2010, https://plus.maths.org/content/florence-nightingale-compassionate-statistician.

243 *"affect thro' the Eyes":* Joshua Hammer, "The Defiance of Florence Nightingale," *Smithsonian Magazine*, accessed July 4, 2022, https://www.smithsonianmag.com/history/the-worlds-most-famous-nurse-florence-nightingale-180974155/.

244 *In 1941:* Antonio Stavrou, Dimitrios Challoumas, and Georgios Dimitrakakis, "Archibald Cochrane (1909–1988): The Father of Evidence-Based Medicine," *Interactive Cardiovascular and Thoracic Surgery* 18, no. 1 (2014): 121–24.

246 *This is best demonstrated:* Ruben Vonderlin et al., "Mindfulness-Based Programs in the Workplace: A Meta-Analysis of Randomized Controlled Trials," *Mindfulness* 11, no. 7 (2020): 1579–98.

246 *Working initially:* Jon Kabat-Zinn, "Mindfulness-Based Interventions in Context: Past, Present, and Future" (2003): 144.

247 *In a similar study:* A. P. Jha et al., "Examining the Protective Effects of Mindfulness Training on Working Memory Capacity and Affective Experience," *Emotion* 10, no. 1 (2010): 54; and E. Denkova et al., "Is Resilience Trainable? An Initial Study Comparing Mindfulness and Relaxation Training in Firefighters," *Psychiatry Research* 285 (2020): 112794.

10. CONDUCTORS

254 *"When a nation"*: R. Kent, "Interview with Dr Randolph Kent—Future Human by Design," Future Human by Design, 2019, accessed July 4, 2022, http://futurehumanbydesign.com/2019 /09/interview-with-dr-randolph-kent/.

259 *"I found the idea"*: Henry Mintzberg, "Covert Leadership: Notes on Managing Professionals," *Harvard Business Review* 76 (1998): 140–48.

260 In 2015: Ture Larsen et al., "A Search for Training of Practising Leadership in Emergency Medicine: A Systematic Review," *Heliyon* 4, no. 11 (2018): e00968; Ture Larsen et al., "Training Residents to Lead Emergency Teams: A Qualitative Review of Barriers, Challenges and Learning Goals," *Heliyon* 4, no. 12 (2018): e01037; and T. Larsen, R. Beier-Holgersen, P. Dieckmann, and D. Østergaard, "Conducting the Emergency Team: A Novel Way to Train the Team-Leader for Emergencies," *Heliyon* 4, no. 9 (2018): e00791.

263 *what follows is the conversation:* "Jeanett Transformation," Ture Larsen, video, 6:54, June 28, 2017, accessed July 4, 2022, https://www.youtube.com/watch?v=GW7XPdnf-EU.

266 *"I had never been"*: Nick Robinson, "I analyse leaders for a living, and none are as great as Alex Ferguson," *Guardian*, May 8, 2013, accessed July 4, 2022, https://www.theguardian.com/football /2013/may/08/alex-ferguson-greatest-living-briton.

266 *And here is Jürgen Klopp:* Paul Joyce, "Jürgen Klopp: A Team is Like an Orchestra—Roberto Firmino Plays 12 Instruments in Ours," Thetimes.co.uk, November 24, 2020, accessed July 4, 2022, https://www.thetimes.co.uk/article/juergen-klopp-football-team-is-like-an-orchestra-roberto -firmino-plays-12-instruments-in-ours-d7vzklwfh.

266 *"just with hand gestures"*: Rosie Pentreath, "What Do Conductors Actually Do? Sir Simon Rattle Has the Answer," Classic FM, February 13, 2020, accessed July 4, 2022, https://www.classicfm .com/artists/sir-simon-rattle/what-do-conductors-actually-do/.

266 *"new spring in their fingers"*: Norman Lebrecht, "Musical Maestros and Football Managers Have More in Common Than You Think," Spectator.co.uk, December 12, 2015, accessed July 4, 2022, https://www.spectator.co.uk/article/musical-maestros-and-football-managers-have-more-in -common-than-you-think.

267 *The first is, in fact:* Quotes from two papers: Melissa C. Dobson and Helena F. Gaunt, "Musical and Social Communication in Expert Orchestral Performance," *Psychology of Music* 43, no. 1 (2015): 24–42; and Gaute S. Schei and Rune Giske, "Shared Situational Awareness in a Professional Soccer Team: An Explorative Analysis of Post-Performance Interviews," *International Journal of Environmental Research and Public Health* 17, no. 24 (2020): 9203.

268 *"Being completely involved"*: Mihaly Csikszentmihalyi, *Beyond Boredom and Anxiety* (San Francisco: Jossey-Bass, 2000).

268 Studies of top-flight: Arnold B. Bakker et al., "Flow and Performance: A Study among Talented Dutch Soccer Players," *Psychology of Sport and Exercise* 12, no. 4 (2011): 442–50; and Christian Swann et al., "A Systematic Review of the Experience, Occurrence, and Controllability of Flow States in Elite Sport," *Psychology of sport and exercise* 13, no. 6 (2012): 807–19.

268 *Research on orchestral players:* Keith Sawyer, *Group Genius: The Creative Power of Collaboration* (New York: Basic Books, 2017).

269 *"Radar [is] a collective"*: Dobson and Gaunt, "Musical and Social Communication," 24–42.

271 *The year 1816:* Robert Evans, "Blast from the Past," *Smithsonian Magazine*, July 2022, accessed July 4, 2022, https://www.smithsonianmag.com/history/blast-from-the-past-65102374; and

Earthobservatory.nasa.gov, "Mount Tambora Volcano, Sumbawa Island, Indonesia," 2009, accessed July 4, 2022, https://earthobservatory.nasa.gov/images/39412/mount-tambora-volcano-sumbawa-island-indonesia.

272 *The widespread loss of horses:* https://www.cyclinguk.org/cycle/draisienne-1817–2017–200-years-cycling-innovation-design; and C. J. McMahon, S. Woods, and R. Weaver, "Sporting Materials: Bicycle Frames," in K. H. Jürgen Buschow et al., *Encyclopedia of Materials: Science and Technology* (Oxford, UK: Pergamon, 2011), 8764–68.

272 *Previously only seen:* Louise Dawson, "How the Bicycle Became a Symbol of Women's Emancipation," *Guardian*, November 4, 2011, accessed July 4, 2022, https://www.theguardian.com/environment/bike-blog/2011/nov/04/bicycle-symbol-womens-emancipation; Adrienne LaFrance, "How the Bicycle Paved the Way for Women's Rights," *Atlantic*, June 26, 2014, accessed July 4, 2022, https://www.theatlantic.com/technology/archive/2014/06/the-technology-craze-of-the-1890s-that-forever-changed-womens-rights/373535/; Worldbicyclerelief.org, "How Women Cycled Their Way to Freedom—World Bicycle Relief," n.d., accessed July 4, 2022, https://worldbicyclerelief.org/how-women-cycled-their-way-to-freedom; Kenna Howat, "Pedaling the Path to Freedom," National Women's History Museum. June 27, 2017, accessed July 4, 2022, https://www.womenshistory.org/articles/pedaling-path-freedom; Jenna E. Fleming, "The Bicycle Boom and Women's Rights," *The Gettysburg Historical Journal* 14, no. 1 (2015): 3; and Sue Macy, *Wheels of Change: How Women Rode the Bicycle to Freedom (With a Few Flat Tires Along the Way)* (National Geographic, 2011).

273 *The great American:* Dawson, 2011.

274 *Her name was Frances Willard:* Ruth Bordin, *Frances Willard: A Biography* (Chapel Hill: University of North Carolina Press, 2014); and Sarah Overbaugh Hallenbeck, "Writing the Bicycle: Women, Rhetoric, and Technology in Late Nineteenth-Century America," Diss., University of North Carolina at Chapel Hill, 2009.

275 *"Simply in learning":* Scot Barnett and Casey Boyle, eds., *Rhetoric, Through Everyday Things* (Tuscaloosa: University of Alabama Press, 2016).

275 *I have an old and worn:* Frances Willard and Carol O'Hare, ed. *How I Learned to Ride the Bicycle: Reflections of an Influential 19th Century Woman* (Sunnyvale, CA: Fair Oaks Publishing, 1991).

277 *Beatrice Grimshaw:* Grimshaworigin.org, *Beatrice Grimshaw, South Pacific Adventurer, Travel Writer and Novelist—Grimshaw Origins and History*, 2001, accessed July 4, 2022, http://grimshaworigin.org/prominent-grimshaw-individuals/beatrice-grimshaw-south-pacific.

EPILOGUE

282 *As Nobel laureate:* Ilya Prigogine, *The End of Certainty* (New York: The Free Press, 2017).

282 *Historian Arnold Toynbee:* Arnold J. Toynbee, *A Study of History: Abridgement of Volumes I–VI* (Oxford: Oxford University Press, 1947).

ABOUT THE AUTHOR

BEN RAMALINGAM is a senior leader, innovator, and researcher specializing in international crisis management and development. He is the executive director of the United Kingdom Humanitarian Innovation Hub and has worked with and advised the United Nations, the World Bank, national and regional governments, nongovernmental organizations, and businesses. In 2020 Ramalingam was named a Humanitarian Change Maker of the 2010s as one of the ten people or organizations who had done the most to improve international humanitarian work over the course of the decade. He is the author of *Aid on the Edge of Chaos: Rethinking International Cooperation in a Complex World*.